Francis de Havilland Hall, Frank Woodbury

Differential Diagnosis

A Manual of the Comparative Semeiology of the More Important Diseases

Francis de Havilland Hall, Frank Woodbury

Differential Diagnosis
A Manual of the Comparative Semeiology of the More Important Diseases

ISBN/EAN: 9783337812287

Printed in Europe, USA, Canada, Australia, Japan

Cover: Foto ©berggeist007 / pixelio.de

More available books at **www.hansebooks.com**

DIFFERENTIAL DIAGNOSIS:

A MANUAL OF THE

COMPARATIVE SEMEIOLOGY

OF THE

MORE IMPORTANT DISEASES.

By F. DE HAVILLAND HALL, M. D.,

ASSISTANT PHYSICIAN TO THE WESTMINSTER HOSPITAL, LONDON.

THIRD AMERICAN EDITION.
Thoroughly Revised and Greatly Enlarged.

EDITED BY FRANK WOODBURY, M. D.,

PROFESSOR OF THERAPEUTICS AND MATERIA MEDICA AND OF CLINICAL MEDICINE IN THE
MEDICO-CHIRURGICAL COLLEGE; ETC.

PUBLISHER'S NOTE

THIRD AMERICAN EDITION.

The present work is founded upon Dr. F. DE HAVILLAND HALL'S *Synopsis of the Diseases of the Larynx, Lungs, and Heart.* The plan adopted by Dr. HALL has, however, been extended to embrace all the more frequent and important diseases.

In the preface to the first American edition the editor stated that he had held especially in view (1) the *early* and often overlooked signs of the presence of disease; (2) the collection of whatever symptoms are alleged on good authority to be *pathognomonic* of pathological conditions; (3) any peculiar features which diseases have been found to present in this country. " Preference has been given to American over European authorities, as every year adds confirmation to the opinion, now widely received, that diseased conditions assume very different aspects under different climatic and sociological surroundings."

In presenting this third edition, the publisher was fortunate enough, as in the second edition, to complete an arrangement with Dr. FRANK WOODBURY to give it a thorough revision, and to add to it what was lacking to make it a complete work within the limits which it aims to cover.

INTRODUCTORY.

In the nomenclature commonly adopted by the best authorities, diseases are divided into two great classes—General and Local. General diseases may be said to comprehend those which affect and pervade the whole system, and in which any local disorder may be regarded as either an accidental complication or sequel; while local diseases are those in which certain organs are especially attacked, and in which the involvement of any other part of the body is considered as consecutive to, or the result of, the primary lesion.

This classification, having much to recommend it, from a clinical standpoint, is the one most practically useful to the physician. The first question he should put to himself on examining a patient is, 'Have we here a general or a local disease? He reaches the answer by excluding those organs whose form and functions present nothing abnormal, and by distinguishing, among such as are implicated, those which indicate primary and essential lesions, from those which are affected accidentally or secondarily. Where no primary lesions are discoverable, he may conclude that he has to do with a general disease.

For the purpose of diagnosis, General Diseases may be conveniently divided into the two morbid groups of (1) Fevers and (2) Diseases of the Blood. These also are each divisible into two or more classes, marked by a few leading and prominent symptoms, which are the guides to the diagnostician. Thus,

The Essential Fevers are usually acute in their course, and either characterized by an eruption of a well-defined character (the Exanthemata); or by a recurrent marked diminution (remission) or periodical total cessation (intermission) of the symptoms (Periodic fevers); or else by a persistent pyretic course not manifesting either of these phenomena (Continued fevers).

Blood Changes are rarely acute, and are either constitutional (the Dyscrasiæ); or else characterized by definite organic lesions (Rheumatism, Gout); or by a physical and generally recognizable change in the blood itself (Anæmia, Leukæmia, Scurvy and Purpura).

Local Diseases are more conveniently classified with reference to the physiological than the anatomical divisions of the body. The functions of life are carried on by the Nervous, Muscular and Osseous systems, and the various organs forming the Respiratory, Circulatory, Digestive and Urino-genital apparatus; and the impairments of each of these form classes of diseases which are broadly discriminated by signs easy of recognition. The niceties of diagnosis are needed rather to distinguish between the varied diseases peculiar to each of these systems than to locate the disturbance in one or the other of them.

CONTENTS.

PART I.
GENERAL DISEASES.

CHAPTER I.
THE FEVERS.

CHAPTER II.
DISEASES OF THE BLOOD.

(vii)

PART II.
LOCAL DISEASES.

CHAPTER I.

DISEASES OF THE NERVOUS SYSTEM.

CHAPTER II.

DISEASES OF THE RESPIRATORY APPARATUS.

CHAPTER III.

DISEASES OF THE CIRCULATORY APPARATUS.

CHAPTER IV.

DISEASES OF THE DIGESTIVE SYSTEM.

CHAPTER V.

DISEASES OF THE URINARY ORGANS.

HALL'S
DIFFERENTIAL DIAGNOSIS.

PART I.

GENERAL DISEASES.

CHAPTER I.

THE FEVERS.

CONTENTS.—*The Febrile State—Inflammatory, or Symptomatic, and Essential Fevers—The Exanthematous or Eruptive Fevers—Typhoid and Typhus Fevers—Typhoid and Malarial Fevers—The Typhoid State—Malarial Fevers—Cerebro-spinal Meningitis—Acute Tubercular Meningitis—Yellow Fever—Relapsing Fever.*

THE FEBRILE STATE.

The most common of all forms of disease is that which is presented by the Febrile State (*Pyrexia*). The chief objective symptoms which it offers are found in

    I. The elevated temperature.

    II. The pulse.

    III. The tongue and throat.

    IV. The urine.

    V. The state of the skin.

    I. The rise of *temperature* is one of the most prominent of the phenomena in fevers, and by many is regarded as the essential feature of the febrile condition; yet its correct appreciation was never understood previous to the labors of WUNDERLICH. Now, the clinical thermometer is considered as necessary to the practitioner as the lancet used to be. For diagnostic purposes, a correct thermometer is now indispensable. It should be self-registering, and should occasionally be compared with a standard thermometer, to insure correctness. The instrument in general use is of glass, with mercurial index. The ther-

2   (17)

mometer should be kept scrupulously clean, but it must never be washed with hot water, or it will be broken.

In using the clinical thermometer, Dr. SYDNEY RINGER, of London, has laid down the rule that in order to insure correctness in the observations, the following conditions must be fulfilled:

1st. That the patient should be in bed, otherwise the temperature of the surface will be much below that of the internal organs.

2d. That the patient be in bed at least one hour before the observations are made, since that time is necessary for the surface of the body to regain the heat lost by previous exposure.

3d. The position of the person examined should be such that the anterior and posterior edges of the axilla are relaxed, for otherwise a cup-shaped cavity is formed, in which the thermometer moves freely without being in contact with its walls. This occurs especially in emaciated persons.

4th. The temperature should be taken twice daily, say at eight in the morning and eight in the evening. If but one observation is possible, then the evening should be preferred, since the morning temperature, abnormal though it may be, rises in the evening.

5th. The thermometer should remain in the axilla at least five minutes.

Although the axilla is generally selected, on account of convenience, the temperature is often taken with the thermometer in the rectum or vagina, especially in children, and it is believed that such observations give more correct indications of the heat of the body than those taken on the surface. Many physicians prefer to place the thermometer in the mouth. The temperature under the tongue is half a degree higher usually than that of the axilla.

The temperature fluctuations in the various zymotic diseases have now been carefully studied by many clinical observers, who have deduced observations which are of great service in diagnosis, as some of them are characteristic.

A pretty constant increase and decrease of temperature exists in the several specific fevers, a close observation of which, in accordance with the foregoing rules, will often serve as a valuable aid both in diagnosis and in prognosis. Dr. WUNDERLICH, in his work, gives useful tables for this purpose, and we subjoin a valuable comparative table of the pulse as well as the temperature in seven of the more frequent febrile diseases, drawn from English observations.

COMPARATIVE TABLE OF THE TEMPERATURE AND PULSE IN THE LEADING FEBRILE DISEASES.

Day.	Typhus Fever.		Typhoid Fever.		Measles.		Scarlet Fever.		Febricula.		Rheumatic Fever.		Pneumonia.	
	T.	P.	T.	P.	T.	P.	T.	P.	T.	P.	T.	P.	T.	P.
1st														
2d			102.	98			104.2	144	103.	99			102.8	123
3d	104.8	108	103.1	98			104.	148	103.7	103			102.3	120
4th.	103.6	113	103.4	110	102.3	130	103.	134	104.	105	101.8	105	103.6	122
5th.	103.	114	102.7	107	103.	124	101.2	122	102.6	99	102.	114	104.	126
6th.	103.2	122	103.2	104	100.2	112	100.6	108	98.4	99	102.	116	103.	122
7th.	104.2	124	103.7	107	98.	102	100.	106			103.	120	102.8	122
8th.	103.8	122	102.5	108	98.	98	100.	110			100.	90	100.	114
9th.	103.	113	103.	108	98.	80	99.8	108			100.	96	99.	94
10th.	102.7	117	102.6	111			99.	100			99.4	86	98.	78
11th.	102.4	119	103.	111			98.6	104			101.	104		
12th.	102.2	108	102.5	112			98.	84			101.	102		
13th.	100.5	106	102.2	108							102.	100		
14th.	100.	100	102.4	109							100.9	100		
15th.	99.4	98	101.8	107							100.	88		
16th.	98.7	92	102.	100							98.	90		
17th.	98.4	90	101.4	100							99.	94		
18th.	98.2	85	98.8	98							102.	96		
19th.			101.4	105							103.	102		
20th.			102.2	100							101.6	100		
21st.			98.8	98							101.7	104		

The above table, prepared from a series of observations, by Dr. J. S. WARTER,* illustrates the normal and average contrasts of pulse and temperature in the course of the diseases specified, when their tendency is to recovery.

II. The *pulse* is increased in frequency, and usually diminished in force; it may also be either hard, full and bounding, or tense, small and contracted. The former condition is more common in active inflammation of the organs above the diaphragm; the latter, in many inflammations below the diaphragm and in idiopathic fevers. In fevers of a typhoid form, an unusually slow pulse is sometimes encountered, and also a pulse with apparently a double beat, the "dicrotic" pulse.

* *St. Bartholomew's Hospital Reports*, vol. ii., p. 78.

In the later stages the pulse may be soft, gaseous or thready, indicating febrile changes in the walls of the vessels and the heart. *Irregularity* in the pulse occurs in pericarditis, and is among the early indications of brain disorder in young children. It is very common in advanced heart disease, but is also met with in functional disorders due to indigestion. *Intermittent* pulse is sometimes met with in typhus and other fevers. A slow pulse is often met with in yellow fever, dengue, and jaundice. A slow pulse also is apt to follow malarial fevers.

III. The *tongue* in the beginning of the febrile state is usually whiter and drier than usual, and more or less coated with a "fur" or viscid covering, from the more rapid evaporation of the watery secretions. Later, in the progress of severe fever, the tongue becomes dry, and the desiccated mucus and epithelium form a brownish or blackish crust, while the papillæ shrink, so that on this crust becoming detached, the surface of the organ looks glazed and smooth. The peculiar appearance of the tongue in certain diseases will be described in connection with these diseases.

IV. The *urine* in fever is scanty and high colored. Its alteration from the healthy average composition is chiefly in the much larger quantity of urea and urates which it contains, and the diminution of its chlorides. According to the researches of Dr. J. BURDON SANDERSON, in the early stage of fever a patient excretes about three times as much urea as he would do on the same diet if he were in health, the difference between the healthy and the fevered body consisting chiefly in this, that whereas the former discharges a quantity of nitrogen equal to that taken in, the latter wastes the store of nitrogen contained in its own tissues. That this disorder of nutrition is an essential constituent of the febrile process is indicated by the fact that it not only accompanies the other phenomena of fever during their whole course, but precedes the earliest symptoms, and follows the latest. That it anticipates the beginning of fever, was first demonstrated by Dr. SYDNEY RINGER in his investigation of the relation between temperature and the discharge of urea, in ague. That the same condition continues after the crisis has passed. *i. e.*, the temperature has begun to sink, was shown by Dr. SQUAREY.

There are various methods of determining the rate of secretion and the amount of urea. Its relative excess may be inferred when the urine has a deep yellow color, a high specific gravity, and a strong urinous odor. If a small quantity of it be allowed to evaporate to a mucilaginous consistence, and nitric acid be added, drop by drop, crystals of nitrate of urea are formed after a few hours. They are of a pearly white lustre, and their proportion roughly indicates the quantity present. When the urea is in great excess, the crystals will form on the addition of nitric acid to the urine, without the preliminary evaporation, by merely allowing the test-tube to stand for a short time. The quantitative determination of urea will be considered under the Diseases of the Urinary Organs.

V. The *skin*, in common with the other emunctories, has its functions much influenced by the fever process. Apart from the changes in temperature, considered above, there are alterations in the appearance of the skin, and in the character of its secretions, which accompany fevers; certain eruptions also appear, that are more or less characteristic; such as the small-pox pustule, the chicken-pox vesicle, the urticarial wheal, and the scarlatinal dermatitis. The parasitic diseases of the skin, although they may accidentally be associated with pyrexia, exist entirely independently, and the fever bears no causal relationship to them. Diphtheria of the cutaneous surface may or may not be accompanied by symptoms of constitutional disturbance.

The skin during fever, has for its typical appearance a color which, if not decidedly dull and sallow, is at least less clear than in health ; in typhus it may be quite dusky; in bilious remittent and yellow fever it becomes jaundiced. In typhoid the surface is more nearly that of health, but the cheeks are flushed; there are also rose-spots on the chest and abdomen. In acute rheumatism, and in the third stage of intermittent, the skin is covered by a profuse perspiration, which in the former case has a sour smell. Exhausting sweats also occur in pyæmia and phthisis. But the dull-colored, dry and harsh skin is the characteristic appearance, and is due to deficient action of the perspiratory and other glands. Certain exhalations of the skin convey contagion, and some of the eruptive disorders have characteristic odors accompanying them, which are exhaled by the skin.

INFLAMMATORY (SYMPTOMATIC) AND ESSENTIAL FEVERS.

The group of symptoms, collectively known as a *fever*, often accompanies strictly local maladies and injuries. In such cases it is distinguished as Inflammatory, or Symptomatic, Fever, and it is of the first importance to distinguish it from Essential, or Idiopathic, Fever, under which general term all true fevers are included. The development of this distinction has been one of the most prominent achievements of the modern methods of diagnosis. "It is astonishing," remarks an eminent writer, " with the progress of medicine, how many affections have been passed over from the domain of fevers to the narrower circle of inflammation of individual organs." Hence it is of prime importance to determine promptly in the beginning of a case whether the febrile symptoms are a feature of a local disease, or the commencement of a general one.

Inflammatory or Symptomatic Fever.	*Essential or Idiopathic Fever.*
Is usually preceded by some local lesions or symptoms.	Has no definite antecedent local symptoms.
Pulse frequent, full and generally tense.	Pulse frequent, full or small, but rarely tense. (DA COSTA.)
Is accompanied by marked and definite local disturbance.	Local disturbances vary, and are not prominent, or but temporarily so.
Course is indefinite, dependent upon the progress of the local lesion.	Runs a definite course, with a strong tendency to spontaneous termination at a given time.
Anatomical lesions marked, definite and invariable.	Generally characterized by obscure, relatively unimportant, or entirely absent anatomical lesions.
Prognosis mainly depends upon the progress of local lesion.	Local manifestations of less importance in estimating the prognosis.

Dr. WILLIAM STOKES* divides the local symptoms of essential fever

* " Lectures on Fever," London, 1874.

into three groups: (1) Functional or nervous; (2) those dependent on special anatomical changes; (3) those arising from re-active inflammation.

Examples of functional symptoms are delirium, carphologia, cough, diarrhœa, epigastric tenderness, and the like; of the second group, the alterations which occur in the brain, heart, lungs, spleen or intestinal glands; and of the third, the swelling and infiltration of organs. What he calls "the grand rule of diagnosis" in fever is *not to apply to these local symptoms in essential fever the rules of diagnosis of local diseases*, as this would lead to a false appreciation of the disease, and to erroneous treatment. For example, a typhus patient may exhibit the marked symptoms of inflammation of the brain; but if he be treated with active antiphlogistic treatment, and with ice to the head, and leeches, he forthwith sinks and dies.

Of hardly less importance is the distinction between *organic* and *functional* (or neurotic) *changes* in fevers. Delirium, pain, coma, convulsions, cough, etc., may all appear as phenomena of the evolution of the poison which produces a general fever, without signifying any definite anatomical lesion. In other words, essential fever produces local symptoms without organic change. " It is," remarks the author just quoted, "because this proposition has not been sufficiently accepted—sufficiently engraved upon the minds of medical men—that so much mischief has been done in the erroneous treatment of fever."

Remarkable cases of inordinate rise of temperature have been reported in hysterical subjects. In a young woman who had accidentally received a concussion of the spine, the temperature of the axilla reached 122° (TEALE). When all possibilities of imposture have been excluded, (such as friction of the bulb of the thermometer), these aberrant cases of high temperature will be found to be very few and far between. The diagnosis from the essential fevers is readily made by attending to the other clinical features of the case in question.

THE DIAGNOSIS OF THE ESSENTIAL OR ERUPTIVE FEVERS (EXANTHEMATA).

This group includes small-pox, varioloid, scarlet fever, measles, roseola, and also those more indefinite forms, varicella and rötheln.

They have many points of similarity. "They are all characterized by a period of incubation, during which the poison lies dormant in the system; by a fever of more or less intensity preceding the eruption; by an eruption which presents a distinct aspect in each disease, and which pursues a definite, clearly defined course, until it, and with it the febrile malady, disappears. Moreover, they are all very prone to occasion serious sequelæ; are all, in the main, disorders of childhood; rarely attack the same person twice; are contagious, and have not as yet been brought under specific treatment." (DA COSTA.)

It is of great credit to the practitioner, and often of the utmost utility to others, for him to make an *early* diagnosis between these diseases. This is not always possible to accomplish. But a close observer will find several indications which will guide him to a correct opinion before the appearance of the rash. Among the principal of these is :—

The condition of the throat. This region is affected at a very early stage in nearly all cases. In simple scarlatina the very earliest symptom is a more or less uniform redness of the middle of the soft palate, the uvula alone, or the uvula, anterior pillars of the fauces, and tonsils; never the posterior wall of the pharynx alone. On the other hand, in small-pox, the part first affected is the posterior wall of the pharynx; while in measles the posterior walls of the fauces and neighboring parts of the pharynx are always redder than the anterior pillars and soft palate (Dr. ALOIS MONTI). In rötheln and measles the tonsils are red and swollen early in the disease; but in simple scarlet fever, for the first twelve hours there is generally very little swelling of the affected parts, and children seldom complain of pain in the neck or in swallowing. After twelve or twenty-four hours the swelling commences, and the redness becomes less uniform, and more punctiform. This peculiar punctiform appearance may be noted often ten or twelve hours before the rash on the skin is visible. Patches of deposit are sometimes seen on the tonsils. If in malignant scarlatinal sore throat, however, there is, from the first, parenchymatous inflammation of the tonsils and the submucous connective tissue, and this condition is associated with well-marked nervous symptoms, a severe case with ulceration of the fauces may be confidently predicted.

In general terms, it may be said that when the soft palate has a diffused red hue, "similar," as TROUSSEAU remarks, "to, but deeper than, that of the skin," while the tonsils are not involved; when with this is a very hot skin, a very quick pulse, vomiting, a tongue with thick, creamy fur, red borders, and prominent papillæ; and with these symptoms, and exposure to the presence of a scarlatinal epidemic, the physician need not hesitate in pronouncing it scarlet fever.

A very early symptom of scarlet fever has been insisted upon as strictly pathognomonic by an Irish physician, Dr. JOSEPH DUGGAN.* It is that the eye assumes a peculiar brilliant and glistening stare, very different from the liquid, tender, watery eye of measles, and which once carefully noted, remains impressed on the observer's memory.

The character of the preliminary fever often differs. In scarlatina it is speedily developed and high, which distinguishes it broadly from diphtheria, which is not marked at the outset; in measles it is of a catarrhal form; while in small-pox it is often associated with very severe pains in the back and loins, not observed in the other exanthemata. This spine-ache is central in its position, and is less affected by change of posture than is the pain of lumbago, and is not confined to one side nor to the *erector spinæ* muscles. It is stated by some authors that this pain is increased in proportion with the severity of the attack and thus forms an important element in the prognosis; but this statement should be confined in its application to adults, as in children the rachialgia is rarely intense. · Dr. WILKS observes that the most virulent cases of variola are almost apyretic and devoid of feverish symptoms, but rapidly sink under the effects of the poison.

In all the exanthemata the eruption makes its appearance in the throat or the mouth, from *twelve to twenty-four hours* (and in many instances longer) before it appears on the cutaneous surface. In *small-pox*, in *scarlet fever*, in *measles*, in all their grades, the eruption may be looked for in this region long before it can be detected at any other point.

Having thus defined the special indications for a diagnosis of these diseases in their earliest stage, we give in the following table a synopsis of their comparative clinical course and phenomena:

* *Medical Press and Circular*, Feb., 1869.

RUBELLA OR RÖTHELN.   SCARLET FEVER.

INCUBATION.

Period of incubation from one to two weeks.

Very uncertain; from less than a day to several weeks; on an average about twelve days.

INVASION.

Languor; shivering; nausea and vomiting; usually sore throat, but not severe. These symptoms may be so slight as to be entirely overlooked.

Shiverings; nausea; vomiting; throat very much inflamed; sneezing and discharge from the nose; convulsions occasionally in children.

Premonitory fever of short duration; relieved by the eruption. No albumen in the urine.

Great heat of skin and very frequent pulse; not relieved by the eruption. Albuminuria very common.

ERUPTION.

Appears early and almost simultaneously over the *whole body*—is *sudden and general*—is less marked on the limbs than on the trunk, and especially on the chest; may first appear upon the back, upon the chest or neck, upon the cheek or upon the forehead; travels downward.

On second day; first on neck, and face, and body; spreads rapidly to limbs.

At first minute dots, which rapidly assume the appearance of large, irregular-shaped patches, somewhat like measles, but less distinct in color and form, varying from three-cent piece to twenty-five-cent piece in size.

The skin feels harsh and rough (like a nutmeg-grater), minute points of redness next appear, soon surrounded by deep rosy areola (like a boiled lobster).

These patches are raised above the surrounding skin, especially toward the middle, and are of a darker red color at the centres.

The eruption is uniform, or in very large patches, of a scarlet hue, with interspersed raised spots and perhaps a few vesicles; the rash is followed, after the seventh day of its appearance, by complete desquamation.

Fades in about four days; desquamation, when it occurs, is fine and bran-like.

The disease is communicable by these epithelial scales.

MEASLES.

SMALL-POX.

INCUBATION.

Generally from seven to fourteen days.

Generally about ten days, but varies from five to twenty days.

INVASION.

Lassitude, shivering, catarrh; sneezing, discharge from nose; harsh cough; rarely vomiting. Conjunctivæ, injected, and eyes watery; more or less photophobia.

Shivering, severe pains in the back; nausea. There may be a marked chill followed by vomiting.

Fever, with hot skin and frequent pulse; rather increased by the eruption.

Fever often very violent, with bounding pulse and pain in the loins; great relief from occurrence of the eruption.

ERUPTION.

Appears on fourth day, first on face, spreads gradually in forty-eight hours to the rest of body.

Eruption at end of third or on fourth day; first on lips, palate and forehead.

Comes out in small, circular dots, like flea-bites. These dots run together and form blotches, of a raspberry color, and the latter are very prone to assume a crescentic or horse-shoe shape, being slightly elevated above surrounding skin.

Eruption is sometimes diffused over the whole body in a confluent form, and is of a dull, deep red color, offering a contrast to the crimson or scarlet redness of scarlet fever. May be very dark in the hemorrhagic form.

Lasts five days; followed by incomplete desquamation.

Eruption is first papular; after about a day becomes vesicular, then pustular; on the eighth day of the eruption, the pustules maturate; about the twelfth the scabs begin to fall. The danger of contagion does not cease until desquamation is entirely completed, and the epithelial scales and crusts destroyed. The vesicles may be discrete or they may run together, as in the confluent variety, and hemorrhages may take place into them in the hemorrhagic form.

RUBELLA OR RÖTHELN.　　　SCARLET FEVER.

ACCOMPANIMENTS.

RUBELLA OR RÖTHELN.	SCARLET FEVER.
Only moderate sore throat, with hoarseness; swelling of the lymphatic glands at the angle of jaw.	Sore throat; coryza or bronchitis rare.
Tongue slightly coated or normal. Cerebral symptoms absent.	Tongue red; raspberry character. Cerebral symptoms frequent.
Moderate systemic disturbance.	Marked systemic disturbance.

THERMOMETRY.

"The temperature always highest on first day of attack, not exceeding 102°, next day falling to 100°, and getting normal on the fifth day." (*Fox.*) "The temperature nearly always *sub-febrile* (99.5° to 100.4°)—sometimes febrile (101.3° to 102.2°)." (*Wunderlich.*)	Temperature may reach 105.6°, or even a higher point. It usually remains continuously high during the eruption, and it is thus "well distinguished from those affections with which, on account of other symptoms, it is most easily confounded, and more particularly *measles and rötheln.*" (*Wunderlich.*) Subsides about the tenth day.
No secondary fever.	No secondary fever.

DESQUAMATION.

Minute particles of cuticle, like scales of *fine bran*. Always begins toward centre of the eruptive patch, and gradually extends to the circumference.	Comes off in branny scales and in large patches. Occasionally epidermis of the hands is detached entire, and may be slipped off like a glove. This is true also of the feet. Itching sometimes excessive.

COMPLICATIONS.

Pneumonia rare.	Pneumonia rare; pleurisy more common. Endocarditis exceptional.

SEQUELÆ.

Dropsy rarely; swelling of the cervical glands not uncommon.	Bright's disease, dropsy, conjunctivitis, deafness, phthisis, chronic diarrhœa, glandular enlargement.
Epidemic, moderately contagious.	Very contagious.

MEASLES.   SMALL-POX.

ACCOMPANIMENTS.

Bonchitis, coryza, and redness of the eyes, constant; sore throat rare.	Sore throat and dry cough; bronchitis rare.
Tongue coated, or red at edges.	Tongue coated and swollen, or red at edges.
Cerebral symptoms very rare and not severe.	Cerebral symptoms, especially convulsions in children, frequent.
Usually mild, but in some cases and in some epidemics, malignant.	Always considerable systemic disturbance.

THERMOMETRY.

Temperature during preliminary fever may reach 105°–106°. Within twelve to twenty-four hours from appearance of rash it sinks speedily to the normal. Protracted defervescence indicates a complication.	Temperature during the preliminary fever high, often 106°; falls rapidly to about 100° after eruption. Rises again during the secondary fever and falls slowly; a slight rise during desiccation.
No secondary fever, though the fever may increase slightly before eruption leaves.	Secondary fever well marked in all cases.

DESQUAMATION.

Always in branny scales, not in patches or flakes.	In scabs, crusts, and thick scales.

COMPLICATIONS.

Catarrhal pneumonia is very frequent, especially in adults.	Pneumonia not very frequent.

SEQUELÆ.

Chronic bronchitis, phthisis, conjunctivitis.	Chronic diarrhœa, glandular enlargement, various diseases of the eyeballs and eyelids.
Contagion usually limited to children. Sometimes exists in epidemic form; soldiers in camp and aboriginal races suffer severely.	Very communicable; mild cases may cause severe or malignant ones. Chiefly adults.

The specific points of diagnosis of epidemic roseola or rubeola from measles are clearly shown in a paper by Dr. I. E. Atkinson, of Baltimore.* His conclusions are as follows:

1. Rötheln (rubella) is a specific contagious eruptive disorder.

2. While it possesses pretty well defined characteristics which, taken together, justify a reasonable degree of certainty in its diagnosis, it has no symptom which may not be, and is not often assumed by measles.

3. A sporadic case, occurring in one who has never had measles, and who affords no history of exposure to rötheln, may be diagnosticated with a fair degree of confidence, but not with absolute certainty.

4. The unqualified diagnosis of rötheln (rubella) should only be made during an epidemic in which all persons exposed, irrespective of former attacks of measles, are liable to be affected, and in whom the symptoms follow a pretty uniform type. In the absence of pronounced epidemic influence, a series of cases occurring in a household, a school, or an asylum, showing typical symptoms, may be diagnosticated as rötheln (rubella) with a fair degree of confidence.

5. In sporadic cases, where neither measles nor rubella has been experienced, a diagnosis of probable measles or rötheln (rubella) must be made, accordingly as the symptoms and course resemble the type of one or the other affection.

In *small-pox* the eruption may often be *felt* before it can be seen, the sensation imparted to the finger being like little shot underneath the skin. Its first appearance is as a simple red point or pimple, soon changing to a papule. The red erythematous blush of *scarlatina* disappears on pressure, but returns when the pressure is removed.

Prof. WM. OSLER, of Philadelphia, and others, have called attention to and described a number of *initial rashes*, which precede, by twelve to twenty-four hours, the appearance of the variolous eruption. They are principally noticeable on the upper part of the trunk, and generally have the similitude of a deep, suffused flush, but vary in their physical characteristics.

The *pulse* in variola is asserted, by some, to be pathognomonic and

---

* In the American Journal of the Medical Sciences, January, 1887, No. 185, page 17.

significant so early in the disease, that the malady can be positively diagnosticated many hours before the eruption appears. But no definite descriptions nor tracings of this pulse have been given.*

TYPHOID AND TYPHUS FEVERS.

Until within comparatively few years, these two fevers were confounded; and although now, in this country at least, they are distinctly recognized as wholly different diseased conditions, yet there are numerous instances where the clinical features of cases assimilate them to one or the other of these conditions, and yet fail to answer satisfactorily their currently-received definitions. Such are the numerous gastric, nervous, simple continued, synochal, mixed, entero-miasmatic, typho-malarial, etc., types which are so often referred to in medical literature. There are, in fact, wide variations in the local features of this group of diseases, and it is the exception to find the classical portraits of one of the group, drawn in the hospital wards of great cities, correspond precisely with the case as seen, modified by the numerous special conditions of particular regions. We shall cite some of these modified types, after having considered the early symptoms and broad distinctions of typhoid and typhus.

Typhoid fever is peculiarly a disease of slow and insidious approach. For days, and sometimes for weeks, the patient is ailing; and as this gradual onset is known to the public, the physician is often called upon to pronounce an opinion as to the probability of the threatenings being of typhoid, long before any positive sign is present.

The general symptoms are a sense of weakness and fatigue, loss of appetite, muscular soreness, headache, (generally dull, sometimes severe,) disturbed sleep, poor appetite, low spirits. A characteristic and often early symptom is epistaxis. Frequently there is a bronchitis, with shallow and rather frequent breathing, with some sonorous râles over the chest. A skilled auscultator can often pronounce from the character of the râles as to the presence of typhoid, as they yield a peculiar, dry, ringing sound. In one of his clinical lectures, Dr. DA

*See Dr. A. S. Payne, *Va. Med. Monthly*, March, 1878; J. S. Conrad, *Trans. of the Med. and Chirurg. Faculty of Md.*, 1874.

COSTA remarks on this: "I should be loath to rest upon this symptom alone, but there is something about it that often makes the diagnosis of typhoid special and specific."* This point is worthy of more notice than it has received.

The pathognomonic symptoms of typhoid are those connected with the abdomen. The belly is swollen and tympanitic; there is usually diarrhœa with perhaps abdominal pains; rumbling near the ilio-cæcal valve and tenderness about the right iliac fossa. The tongue is coated, and sometimes moved with pain; the teeth show accumulations of dried mucus (sordes); thirst is rarely excessive; vomiting is rare; the mind is dull, and the delirium is usually low and muttering. The peculiar eruption appears on the chest and belly, most frequently between the nipple and navel, about the sixth or eighth (never before the fifth) day of the fever. It is in scattered, small, reddish, deleble spots, resembling flea-bites, which come out in successive crops. Later they become rose-colored, and are surrounded by an area of erythema, shading off into the surrounding skin. They are not elevated, or very slightly so, and they disappear entirely on firm pressure, but promptly re-appear. They give no feeling of hardness to the finger passed over them. These spots may appear all over the body, but are often wholly absent however; and their presence, number and size do not seem to bear any relation to the severity of the attack. Sudamina are also frequently seen, but are not considered of much import.

The prodromal symptoms above mentioned are, however, often varied. Dr. A. LARRABEE, of Louisville, remarks that the characteristics of the prodromal stage, the lassitude, epistaxis, and even the susceptibility of the bowels to purgatives, which are valuable aids to early diagnosis in more northern latitudes, are not so important in the malarial regions of the southern and southwestern United States;† and Dr. JURGENSEN, of Kiel, Prussia, has given the history of a number of cases, with the anatomical characteristics of typhoid fever, when the attacks were sudden, with a well-marked chill, a high temperature (104° Fahr.) and quick pulse, swelling of spleen and little or no diarrhœa.‡ Such a course is extremely rare in this country.

*Medical and Surgical Reporter, Vol. xxviii, p. 11.
† Trans. Kentucky State Med. Soc., 1876, p. 123.
‡ Med. Times and Gazette, 1874.

In *typhus* the eruption usually appears as small, discrete spots, slightly elevated, of a dingy red color and not completely fading on pressure. In a short time the spots cease to be elevated, and fade less on pressure, and a purple mottling appears in the interjacent portions of skin. At a still later period—say on the eighth, ninth or tenth day —the spots become entirely petechial, not being at all affected by pressure. The eruption begins to fade about the ninth or tenth day, and disappears about the fourteenth, and, if there be no local complication and the patient has not been very greatly prostrated, convalescence is established between that day and the twenty-first. In slighter cases, however, the serious illness may continue for only about a week, the eruption may never be very marked, and the patient may become convalescent from the tenth to the fourteenth day; while in very severe cases the rash may become petechial at an early period, and may continue on the skin till near the end of the third week, and the convalescence may be very greatly protracted. Generally in a simple uncomplicated case of typhus, the pulse and temperature fall below the normal standard at the earlier period of convalescence, and again rise when the patient takes more food and is capable of some little muscular exertion. Usually the bowels are confined during the course ·of the disease, but such is not always the case; sometimes towards the height of the fever and when there is great prostration of strength, the bowels are relaxed, apparently from want of power in the sphincter to retain the fecal matter; but in some cases there is profuse diarrhœa during the whole course of the disease, and this independently of any cathartic medicine. It is impossible to base the diagnosis between typhus and typhoid upon the confined state of the bowels in the former disease and the occurrence of relaxation in the latter, for it is not uncommon for the bowels to be confined in typhoid. So, also, though the cerebral disturbance is more marked in cases of typhus than in typhoid, as the rule, it sometimes happens that a patient will pass through a marked or even severe attack of typhus without much delirium, retaining his intelligence to such an extent as to be able to answer simple questions put to him, without any apparent difficulty. In such cases, however, the patient, on recovery, usually has no recollection of anything that has occurred from a very early period of his illness till convalescence is far advanced.

3

In the following table, the leading phenomena of the two diseases, as given by the best authorities, are contrasted, for the purpose of establishing their clinical distinction. The non-identity of the two affections is now everywhere acknowledged. (In Germany typhoid is often called abdominal typhus—an unfortunate title.)

TYPHOID. (ENTERIC FEVER.)	TYPHUS. (SHIP FEVER.)
Age generally from eighteen to thirty-five.	At all ages, often in persons beyond middle life.
Only very exceptionally contagious; often sporadic or endemic.	Highly contagious, generally epidemic.
Attack generally insidious in its approaches.	Attack generally sudden; no lengthened prodromata.
Duration usually fully three weeks; often much longer.	Duration somewhat shorter; often not prolonged beyond second week.
Death hardly ever occurs before end of second week; generally in or after third week.	Death not unfrequently occurs at end of first week, and often before conclusion of second.
Cerebral symptoms come on gradually; last longer.	Delirium or decided stupor comes on soon, sometimes almost from the onset; headache has appeared and disappeared by about the tenth day.
Great emaciation.	Less emaciation; greater prostration.
Face pale, or flush confined to cheeks.	Face deeply flushed, of dusky hue; eye injected.
Skin hot, sometimes covered with acid perspiration.	Skin of pungent heat, sometimes emitting an ammoniacal odor.
Abdominal symptoms, such as diarrhœa, tympanites; intestinal hemorrhage not unusual.	No abdominal symptoms; bowels constipated; meteorism rare; no intestinal hemorrhage; sometimes acute dysentery during convalescence.
Epistaxis common.	No epistaxis.
Bronchitis and sometimes pleurisy.	Pneumonia or marked congestion of the lungs, and bronchitis of finer tubes.

TYPHOID.
(ENTERIC FEVER.)

Eruption light red and not usually on extremities; it is discrete.

Autopsy shows morbid state of Peyer's patches and solitary glands; enlargement of mesenteric glands; ulceration of mucous coats of intestines; enlargement and softening of spleen; ulceration of the pharynx.

TYPHUS.
(SHIP FEVER.)

Eruption of a darker color and all over the body.

No constant *post-mortem* appearances; most common are dark-colored, liquid blood, and enlargement of the spleen; softening of heart more common than in typhoid; no intestinal lesions.

COMPARATIVE THERMOMETRY (DR. J. W. MILLER).

TYPHOID.

The duration of elevated temperature is very rarely less than twenty-one days; it is generally longer, and may be protracted to thirty-five days, or even more.

The evening temperature is almost constantly higher than that of the morning.

The difference between the morning and evening temperature is generally, throughout the case, greater than in typhus; and toward the end of the fever there occurs the very characteristic oscillation of temperature, during which the difference is frequently five, six, or even seven degrees, and which may continue from a few days to a week or more.

A high temperature is frequently accompanied by a pulse but slightly accelerated, and occasionally by a pulse slower than normal, and not infrequently dicrotic, especially during convalescence.

TYPHUS.

The duration of elevated temperature is very rarely beyond eighteen days; it is generally shorter by several days, and may be even so short as nine days.

The evening temperature is frequently lower than that of the morning.

The difference between the morning and evening temperature, during the height of the fever, or from about the third to the tenth or eleventh day, is comparatively seldom above one degree, and although about the period of defervescence the difference is sometimes much greater, the oscillation is not continued over more than one or two days.

A high temperature is, as a rule, accompanied by a high pulse.

The varieties of fever called *gastric* and *nervous* have not been recognized as distinct types by the most recent writers. Yet there is no doubt that many cases of continuing fever present gastric rather than abdominal symptoms, and various other perceptible variants from the type of a mild typhoid. The following semeiological table, drawn from Dr. COPLAND's work, will illustrate this :—

FORMS OF TYPHOID.

SIMPLE CONTINUED TYPE.		NERVOUS TYPE.
100–120, small, weak, irregular; intermittent when a dangerous attack.	Pulse.	Soft, feeble, and quick; about eleventh day very quick and unequal.
Heat of surface generally rises over 100°.	Temperature.	Heat of skin not much elevated; it may even seem natural or diminished.
White, foul, loaded or furred; again red at its sides, and point loaded with dirty yellow fur.	Tongue.	Loaded or covered with a dirty mucus, afterward brown or black, incrusted or fissured.
Tenderness at epigastrium; looseness or diarrhœa of an ochery hue; vomiting early.	Gastric Symptoms.	Fetor of the breath and of the discharges, an irregular relaxed state of the bowels, pain at the epigastrium, nausea and vomiting.
Pain in head, throbbing of arteries, brilliant expression of eyes, marked acuteness of senses, watchfulness and restlessness, moaning and incoherent muttering, dilated pupils, and coma.	Head Symptoms.	Countenance pallid or transiently flushed, head heavy, continual restlessness, want of sleep, tremor, hearing dull, unconscious evacuations, low delirium, early stupor and coma.
A common and early complication, either of bronchial surface or congestion of substance.	Lung Symptoms.	The bronchial surface is the part chiefly affected; substance of the lungs sometimes complicated.
Sore throat or inflammation of fauces sometimes accompany.	Affection of Throat.	Sore throat, occasionally so severe as to resemble an attack of anginosa maligna.
By subsidence of the prominent morbid actions indicative of a gradual decline.	Recovery.	Often announced by a true crisis.

Gastric fever, which is not to be confounded with gastritis (although this is more properly a fever of gastric origin), is recognized by NIEMEYER and other competent authorities as a separate type. It commences with loss of appetite, headache and languor, followed by a slight chill, with marked gastric irritability, great nausea, frequent vomiting, and constipation. There is considerable tenderness on pressure over the stomach, a low pulse (60 to 70 per minute), and a temperature at first rising slightly to (100° Fahr.), then falling below the normal as the disease advances (to 95° and even lower). A grave symptom is double vision or total loss of sight. There are no tympanites, diarrhœa, delirium, subsultus tendinum, spots, iliac tenderness, nor sordes, as in typhoid. Women are more liable to it than men, old persons than those of middle age or youth. Its outbreaks indicate it to be a zymotic disease, and the mortality is even higher than in ordinary typhoid. The pathognomonic symptom of the disease is the peculiar sweetish odor of the breath; it is likened by some to the odor arising from hot water poured on garlic, having a slightly alliaceous odor; or, according to others, it.resembles a faint aroma of valerianic acid.* What is usually called gastric fever is really typhoid.

Typhlitis can readily be distinguished from typhoid fever by the pathognomonic sign of a dense tumor in the iliac fossa, increasing, and exceedingly tender on pressure.

TYPHOID AND MALARIAL FEVERS.

TYPHO-MALARIAL (Woodward), ENTERO-MIASMATIC (Wood), OR REMITTO-TYPHUS (Drake).

In order to bring into relief the broad distinctions between the typhoid and malarial fevers when in their typical forms, the following comparative table has been prepared, which is chiefly that of Dr. E. M. Hume.†

TYPHOID.		MALARIAL.
Decomposing animal and vegetable matter, especially human excrement.	Cause.	Emanations from marshes, damp, low, or new soil; always vegetable, never animal.
Old soil; may be high and dry and long settled, especially where saturated with sewage.	Locality.	New land, moist, low and swampy.
Epidemic of typhoid fever.	Circumstantial Evidence.	Prevalence of malarial disease.
Seldom after forty.	Age.	All ages.

* Dr. G. B. Bullard, *Trans. of the Vt. Med. Soc.*, 1877, pp. 52-56.
† *Peninsular Journal of Medicine*, Feb., 1875.

TYPHOID.		MALARIAL.
Continued without intermission or decided remissions.	Periodicity.	There is either intermission or remission.
Lasts three or four weeks; cannot be cut short.	Duration.	Can be interrupted and cured in a few days.
Great nervous disturbance and prostration; dull, heavy, throbbing, persistent frontal herdache; twitching of muscles; tickling of throat; ringing in ears; deafness; mind stupid.	Nervous Implication.	None, although there is sometimes severe headache, simulating meningitis.
Asthenic, not wild.	Delirium.	Sthenic.
Frequent.	Epistaxis.	None.
Catarrhal bronchitis with sometimes tough, tenacious sputa.	Lungs.	Congested, when affected at all.
From 70 to 140 beats per minute, small, irregular or dicrotic.	Pulse.	More frequently high, full and bounding.
Hot, even when moist; emits a peculiar, musty odor, pathognomonic of this fever.	Skin.	Varies; sometimes dry and hot; odor acid and swampy; at other times may be normal.
Indicates an increase of temperature from morning to evening of about 2 degrees, and a decrease of 1 degree from night to morning; commences first day 98.5 degrees, reaches its maximum of 104 degrees on the morning of the fourth day; from this time the evening temperature ranges between 103 degrees and 104 degrees, morning 1 degree lower, until end of second week, when it gradually declines in the same regular manner, always lower in the morning than in the evening, except when there is a complication.	Thermometer.	Rises rapidly to 105 degrees or more first day or two, and falls suddenly; is not so uuiform; may rise and fall seven degrees in one day.
Protrudes tremblingly; is covered with a whitish yellow coat, which subsequently disappears and is replaced by a dry, pale brown one, with red, glazed tip and edges; teeth covered with dark brown sordes.	Tongue. .	Coated all over with a heavy, dark-yellow coat. No sordes.
Pale, livid, muddy, or may be clear, with cheeks flushed.	Complexion.	Sallow; eyes yellow.
Foaming, light color, free from sediment; frequently contains albumen; has typhoid odor like body.	Urine.	Dark color, turbid, no albumen, except in malarial hemorrhagic fever.

TYPHOID.		MALARIAL.
Diarrhœa, except in mildest cases; stools offensive, pea soup, bright yellow or brown; devoid of mucus, but sometimes contains whitish flocculi.	Excretions from Bowels.	Bowels costive; dark, hard, dry, bilious stools.
Tympanites occurs, giving tub shape to abdomen; pressure over cæcum produces pain and gurgling sound; tenderness over spleen.	Abdomen, Shape, etc.	No tympanites or tenderness of abdomen.
Stomach not involved; no severe pain anywhere, except when peritonitis occurs.	Pain.	Gastric disturbance and vomiting of bile; pain in stomach and elsewhere very intense; may be throughout the entire body.
Occurs during second week; from one to twenty small rose-colored spots, size of pin head, appear on abdomen, chest or back; do not extend to extremities; present no distinct elevation to the touch, disappearing upon pressure, but reappearing upon its removal; last about three days, fade away, and a fresh crop appears. This eruption is claimed to be "peculiarly and absolutely diagnostic of typhoid fever." Later in the disease sudamina appear.	Eruption.	Eruptions of different kinds sometimes occur, but are so different in shape, feel, duration, number, extent and place, that they need never be mistaken for the typhoid eruption.
Great—averages one in five.	Mortality.	Very slight—not one fatal case in a hundred.
Inflammation and ulceration of Peyer's, solitary and Brunner's glands; sometimes perforation of bowels, with peritonitis, and fatal hemorrhage; inflammation and enlargement of mesenteric glands and the spleen (which sometimes bursts); the brain, stomach, liver and lungs sometimes inflamed.	Lesions.	Hemorrhage from congestion of bowels rare; congestion of stomach, lungs, liver and spleen—the two latter sometimes become enlarged.

We shall now consider the character of a combined form of disease presenting in its different stages symptoms both of malarial and typhoid fever.

The experience of numerous observers has proven that there is a complex form of fever prevalent in malarious districts, in which the typhoid and miasmatic elements are combined. It has been proposed by Dr. J. J. WOODWARD to call this "typho-malarial fever," a term which he explains to be applied "not to a specific or distinct type of disease, but to the compound forms of fever which result from the combined influence of the causes of the malarious fevers and of typhoid fever." *

The name *Remitto-Typhus* was given to it by Dr. D. DRAKE, who also spoke of it as " the typhoid stage of remittent or autumnal fever." He does not consider it a distinct disease, but a genuine hybrid of typhoid and remittent fevers. He remarks that in many cases the stage of invasion is of nearly the same length in both; both attack males more than females; and that when remittent terminates fatally, subsultus tendinum, a dry tongue and intestinal hemorrhage are sometimes present. He has, however, never seen a decided intermittent pass into typhoid; nor a well-marked typhoid terminate in an intermittent.†

During and since the war, typho-malarial fever has attracted much attention, and its traits have been thus distinguished from simple typhoid.

TYPHOID.	TYPHO–MALARIAL.
Occurs in all localities, most common in the north.	Only in miasmatic localities; most common in the south.
Invasion gradual and without remittence.	Often begins as simple intermittent or remittent.
Daily exacerbation and remission slight.	Decidedly marked.
Diarrhœa the rule. Tympanites common. Abdominal tenderness considerable; epigastric and hepatic tenderness slight.	Constipation the rule. Tympanites rare. Abdominal tenderness slight; epigastric and hepatic tenderness considerable.
Temperature comparatively low. Delirium low and muttering.	Temperature high, especially at outset. Delirium active.
Spleen not involved to the same extent.	Tumefaction of spleen very marked.
Sordes on the teeth the rule.	Sordes rare.
Peyer's glands always involved.	Rarely involved.
Rose-colored eruption present.	Generally entirely absent.
Pigment deposits absent.	Pigment deposits in various tissues and organs very common.

* *Transactions of the International Medical Congress*, 1876, p. 340.
† "Diseases of the Interior Valley of North America," p. 556.

THE TYPHOID STATE.

It is a common error to confound the typhoid condition, which occurs in many diseases, with typhoid fever, properly so-called. This typhoid state may be developed in typhus and other fevers, in acute pneumonia, rheumatism, tuberculosis, pyæmia, and various renal diseases, epecially the granular or gouty kidney, and Bright's disease. The exciting cause in all these cases, it is believed, is the accumulation in the blood of the nitrogenous products of disintegration of the tissues.

The so-called "typhoid symptoms" are a quick, soft pulse; a dry, brown tongue; the phenomena and physical signs of hypostatic congestion of the lungs; impairment of the mental faculties; stupor passing into coma; delirium, which is at one time acute and noisy, at another low and muttering, and not infrequently associated with muscular tremor; involuntary discharges. The skin is dusky, moist and often emitting a fetid odor. There is little thirst and often difficulty in swallowing. The temperature and urine vary considerably. The respirations are shallow and somewhat accelerated. The bowels are sometimes constipated, but often relaxed with offensive evacuations.

The difference between this condition, as it supervenes in the above named diseases, and the true typhoid, or continued fever, may be thus presented:

THE TYPHOID STATE.	TYPHOID FEVER.
Arises in the course of antecedent disease.	Begins insidiously without any history of preceding disease.
Is always traceable to debility or blood-poisoning from deficient elimination.	Can often be traced to an external zymotic or septic influence.
Abdominal symptoms generally absent.	Diarrhœa, tympanites, epistaxis, tenderness over intestinal glands, pain in iliac fossa, as the rule.
Occasionally there may be papules or spots of diffused rosiness, from dilatation of superficial capillaries, but nothing like the *tâches rouges*.	Eruption of rose-colored spots, coming out in crops at the end of the first week.

THE TYPHOID STATE.	TYPHOID FEVER.
Intestinal hemorrhage not to be expected.	Intestinal hemorrhage not infrequent.
Splenic enlargement not usual, except in malarious cachexia (ague cake).	Enlargement of spleen very constant.
Urine may show albumen or pus.	Pus not present, albuminuria occurs rarely.

MALARIAL FEVERS (PALUDISM).

The characteristic symptom of all malarial affections is *periodicity*. It is not, however, pathognomic; for hectic and syphilitic fevers, neuralgia and many other disorders, simulate this trait very closely. The diagnosis, however, in most instances is not difficult except when poisoning by malaria complicates other diseases.

Intermittent fever begins with a chill, cold extremities, pale face, chattering teeth and feeble pulse, followed by a decided fever, in which the face becomes flushed, the skin hot, the pulse full and rapid; and it ends with a profuse perspiration, soft, moderate pulse, and restoration of the secretions. This occurs at regular, definite intervals, with complete intermissions between.

In *remittent fever* we find the same development of the phenomena, the chill, the fever, the perspiration, but without complete abatement of the febrile symptoms in the interval. They continue, though lessened, and usually have daily exacerbations. It is generally preceded by intermittent.

Between these two most common forms there are the differences that in intermittent fever the patient is well between the paroxysms, in remittent he continues ailing; in intermittent a distinct chill precedes each attack, in remittent the chills are slight or absent; in intermittent the appetite is good between the invasions, in remittent nausea and anorexia are present. Dr. DANIEL DRAKE says: "If we suppose an ague shake to be reduced to a mere chill, but the subsequent hot stage aggravated and prolonged, we shall form a just conception of the relations, in symptomatology, between intermittent and remittent fever."*

* "Diseases of the Interior Valley of North America," p. 95.

The more intense cases of malarial poisoning develop *algid, pernicious, or congestive chills, malignant, remittent fever*, and *malarial hemorrhagic fever*.

In *congestive chill* the symptoms of an ordinary intermittent are present, but in an exaggerated form. The chill is intense, the skin and even the breath seem cold; the face is cadaveric; the respiration is sighing; the pulse scarcely distinguishable; the shivering shakes the bed. The patient may die in the chill of internal congestion. When the stage of fever comes on, the pulse is full, and so quick it can scarcely be counted; the skin of the body is hot while the feet and hands are cold; delirium is active; thirst intense; the stomach is irritable. The perspiration that follows brings no relief; the patient lies prostrate and sometimes unconscious or comatose. When the congestion affects the lung there is an air-hunger, difficult breathing, and bloody expectoration; when it attacks the stomach and bowels there are violent spells of vomiting, foaming or soap-like white discharges, and great epigastric tenderness. In these cases the mind is usually clear; but when the brain is involved, there is intense headache, the mind is dull or delirious, and coma is apt to supervene. Patients may die in the first or second, and but rarely survive the third chill of this intensity.

The diagnosis of *malignant remittent* has been carefully set forth by Dr. DANIEL DRAKE as follows:—

1. The pulse does not rise in fulness and force during the exacerbation, as in other forms of remittent fever, but is generally small, frequent, weak, and variable. When the remission begins it generally improves slightly, but to a much less extent than in mild remittents.

2. The feeling of abdominal oppression, and the anxiety, restlessness, and gastric irritability are deeper, in this than in other forms of remittent fever; and these symptoms never entirely cease during the remission.

3. A coldness in the hands and feet, or of the ends of the toes and fingers only, continues through the hot stage, while the trunk of the body and the head are in high fever heat. With the arrival of the remission, this coldness, in milder cases, is replaced by a natural temperature; but in the more malignant it continues. Many experienced

physicians regard this as the most characteristic sign of malignant remittent.

4. There is no time when the fever is absent; and whatever irrita-
tions or congestions are formed in the cold stage, and whatever inflammations are set up in the hot stage, remain, though moderated in degree, throughout the remission.*

Hemorrhagic malarial fever commences with a chill of the congestive type; and during the first paroxysms the symptoms which distinguish this from all other fevers usually make their appearance. These
are jaundiced skin and vomiting, apparently without any effort, of a
dark fluid; the fæces dark, offensive, and tawny looking; the color of
the skin yellowish or bronzed, and the *urine colored with hæmaglobin.*
The last-mentioned is pathognomonic. Sometimes, though mixed
with blood, the urine is profuse, which is a favorable symptom. Most
of such cases recover; but when the urine grows scanty, and suppression ensues, the result is said to be always fatal.† The remissions are
irregular and often ill-defined ; and after the hot stage it has been noticed that there is no perspiration.‡ Pain in the back is severe and
incessant; the stomach is irritable, and the mental powers often ob-
scured.

A characteristic color of the tongue in malarial poisoning has been
observed by Professor CHARLES O. CURTMAN, M. D., of St. Louis. He
describes it as almost uniformly present. The color of the dorsum of
the tongue as far back as the circumvallate papillæ is of a bluish-gray
tinge, somewhat resembling that of old sheet zinc or lead. It occurs
in various degrees of intensity, giving the impression of a coloring of
greater or less thickness, superimposed upon the epithelial surface,
sometimes quite thin and transparent, at other times quite opaque. In
some cases this hue is observed without any other pronounced symptoms of malaria; but in all such the distinct malarial symptoms follow.
The disappearance of this color serves as a valuable index of the per-
fect restoration to health.||

* *Loc. cit.*
† Dr. GREENSVILLE DOWELL, " Yellow Fever and Malarial Fever," p. 213.
‡ Dr. THACKER, Cincinnati *Medical News,* 1872.
|| St. Louis *Medical and Surgical Journal,* 1869.

The symptoms of malarial poisoning are multiform, and are frequently so masked and disguised that the closest observation fails to detect their origin. The entire organism may be affected by the poison. This is the condition of *malarial toxæmia.* It is broadly characterized by a tendency to cerebral, thoracic and abdominal congestion, obstinate to ordinary remedies, and often slightly but distinctly periodic in exacerbations. Bronchitis, diarrhœa, simple fever, toothache, neuralgia, ophthalmia, urticaria, and other skin diseases, even hæmoptysis, hysteria and rheumatism, may all be caused, instituted, or simulated by this subtle poison in a community subject to its influence.

Careful examination will usually disclose evidence of periodicity in an increase of suffering in these cases at regular periods; sometimes at intervals of several days, or even weeks apart; or they may be regularly aggravated at morning, noon, or night. Subordinated to the prominent symptoms, and apt to be overlooked by the patient unless particular inquiry is made, are slight recurrent headaches, intolerance of light, shiverings, or a sense of cold, or alternating heat and cold, or perspirations. A trace of blood in the urine, especially in the tropics, is a common indication of malaria. Nausea or vomiting, or a copious watery discharge from the bowels at periodic intervals, are often observed, especially in children.* The skin is harsh, dry, and presents a muddy or else a greenish-yellow hue, which is most noticeable on the face, neck, and arms. The appetite is capricious, the strength easily exhausted, the temper irritable, the mind readily depressed, and the energies diminished. On careful percussion the spleen is nearly always found to be decidedly, and the liver slightly, larger than in health.

The condition of the blood in malarial poisoning has been studied with definite results. Dr. A. KELSCH has found that the white corpuscles diminish during an attack to one-half or one-third of their normal number, and continue less than usual so long as there is splenic enlargement.† Malarial anæmia is occasioned by destruction and decolorization of the red corpuscles.

* See an article on "Infantile Malarial Toxæmia," by Dr. JOEL C. HALL, in the *Medical and Surgical Reporter*, Vol. xxxi., p. 147.

† *Archives de Physiologie*, October, 1876.

Various observers have reported the presence of characteristic appearances of the blood in addition to the granular pigment masses derived from the red blood corpuscles. Salisbury described the toxic agent as palmella, Klebs as a bacillus, Moss and others, bacteria singly, in pairs, or in zoöglea groups. The observations of Laveran seem conclusive, however, in demonstrating the existence in the blood of an organism belonging to the flagellate infusoria. Marchiafava and Celli have confirmed these observations, and propose the name of *plasmodium malariæ* for the newly-discovered organism. Councilman, of Baltimore, and Osler, of Philadelphia, have also found these bodies constantly present, and urge their importance in diagnosis, without, however, absolutely committing themselves to the opinion of their causal relation with the disease. Osler describes * the bodies as occurring both inside the red corpuscles, and free in the plasma. The intra-cellular form appears as either a hyaline or a darkly-pigmented body, filling one-third or one-half of the corpuscle, and undergoes slow amœboid changes. The hæmoglobin of the corpuscle is gradually destroyed by the organism, and the stroma becomes pale and finally colorless. There seems to be no doubt whatever about the amœboid character of these movements, which are readily followed with a high-power objective. The forms occurring outside the corpuscle are still more remarkable. These are: (1) Small, circular, pigmented bodies; (2) Curious crescent-shaped organisms; and (3) An extraordinary flagellate body resembling an infusorian. The pigmented crescents have been noted by all observers, and are much more readily seen than the amœboid bodies. They do not occur so frequently, and apparently only in the later stages of the disease. The flagellate form, also pigmented, is still less common. The movement of the flagella is very active, so that it brushes away the red corpuscles in its vicinity. The relation of these forms to one another is still doubtful, though they probably represent phases of development of the same organism, which is not a bacterium, but is classed with more propriety among the monads.

* *Proceedings Philadelphia Pathological Society;* also editorial in the *Philadelphia Medical Times*, November 13, 1886.

CEREBRO–SPINAL FEVER (EPIDEMIC MENINGITIS, OR SPOTTED FEVER).

This disease is very apt to occur in epidemic form, as is expressed in one of its titles. Its onset is usually sudden, often beginning with a chill, vomiting and intense headache, and an elevation of pulse and temperature. The pathognomonic symptom is that *the head is drawn backward and downward, and the muscles at the back of the neck are rigidly contracted, very tender to the touch and painful on motion.* The pupils are also contracted.

At an early period herpes may appear on the face or limbs, the skin is hyperæsthetic, and the patient cannot bear handling. After about four days convulsions may set in, or tetanic contractions make their appearance, and stupor follows, passing into a coma, preceding disso-lution. The bowels are usually persistently constipated, and the urine passes involuntarily.

In cases tending toward recovery the acute symptoms gradually subside, and, after a week or two, convalescence takes place, attended by more or less headache and muscular contraction.

In regard to differential diagnosis, it may be simulated by typhus or masked variola. The absence of tonic spasm of the post-cervical muscles in these diseases will aid in recognizing them. The pro-tracted cases, where this symptom is not prominent, may resemble typhoid fever. In both there is an eruption, some similar cerebral symptoms, and occasionally intercurrent diarrhœa. But the invasion of cerebro-spinal meningitis is more sudden, the headache more violent, and there is vomiting and constipation; while later the spinal pain, the herpes, the tetanic spasms and the continued headache, are broad distinctions.

True tetanus is distinguished by the absence of epidemic prevalence, by the clearness of the mental powers, and by the history of the case pointing to some injury.

Certain forms of malignant malarial fever counterfeit cerebro-spinal meningitis, especially during convalesence, when the affection presents periodical intermissions of the febrile state. The points of difference may be summed up as follows (HAMILTON):

CEREBRO-SPINAL MENIN-GITIS.	CONGESTIVE PERNICIOUS MALARIAL FEVER.
Inceptive chill not marked.	Chill quite marked.
Disease epidemic, and chiefly among children.	Endemic and common to all ages.
Muscular spasms the rule.	Muscular spasms very rare.
Bowels constipated.	Not usually so.
Pulse and temperature do not suffer rapid variations.	Both subject to great variations, feeble and irregular.
Temperature does not undergo periodical changes.	Undergoes decided periodical changes.
Face flushed; eruption before fourth day.	Complexion sallow; no eruption.
Delirium and coma not affected by large doses of quinine.	All symptoms modified usually by large doses of quinine.
Increase of fibrin and rapid coagulation of blood when drawn.	Malarial organisms can be detected in the blood.

In distinguishing it from other head affections it should be observed that, while pain in the head, vomiting, epileptiform attacks, disease of the optic discs, emaciation, eruptions, involuntary micturition, are symptoms found in many of them, the sudden onset of fever, pain in the back of the neck, the stiffness of the muscles of the neck, and retraction of the head, are sufficient to separate cerebro-spinal meningitis from *hydrocephalus acquisitus, basilar meningitis, and tumor of the brain*, diseases to which, in its symptoms, it is nearly allied.

It may also be noted that Dr. HAYDEN, of Dublin, a competent authority, states that he never saw a case of cerebro-spinal meningitis unattended by *pains in the calves of the legs*, and he should make a presumptive diagnosis from the presence of that symptom alone.

Dr. DOWSE, of London, has insisted on the importance of distinguishing *sporadic* from *epidemic* cerebro-spinal meningitis. He maintains that in its epidemic form the sensorium is more or less affected from the first, and the membranes over the superior cerebral convolutions, cerebellum, and posterior columns of the cord, including the nerve substance, are primarily, if not wholly, the seats of lesion. In the sporadic form, on the contrary, the sensorium and special senses are only slightly influenced, and the inflammation centres itself upon the meninges at the base of the brain and the anterior columns of the cord. He therefore gives to the latter affection the name of *occipital* or *basic cerebro-spinal meningitis*, in contradistinction to the former well-known disease. He draws his conclusions and diagnosis from signs and symptoms, as evidenced in the following table:

EPIDEMIC CEREBRO-SPINAL-MENINGITIS.	SPORADIC OR BASIC CEREBRO-SPINAL MENINGITIS.
Attack sudden, without any special predisposing cause.	Attack commences gradually and resembles an onset of acute rheumatism.
Apparently of a contagious or infectious origin.	Usually arises from exposure to cold, exhaustion, and privation.
Sensorium affected from the first.	Sensorium never affected until the last stage.
Excito-motor spasms of a tonic character in groups or groupings of muscles, with marked loss of cutaneous and muscular sense.	Incoördination of movement, with cutaneous formication, partial anæsthesia, muscular hyperalgia, but no tetanic spasms.
Reflex movements common.	Reflex movements rare.
Vomiting urgent and uncontrollable.	Vomiting not so severe.
Temperature rarely exceeds 100°.	Temperature often rises to 105°.
Purpuric maculæ diffuse and general.	Maculæ never seen in the desudate form.
Death usually takes place from coma.	Death usually takes place from apnœa or exhaustion.
Prognosis grave.	Prognosis hopeful; much affected by treatment.
Post-mortem appearances reveal the membranes over the superior cerebral convolutions and posterior columns of the cord as the seat of lesion.	Post-mortem appearances reveal the membranes over the base of the brain and over the anterior columns of the cord as the prime seat of lesion.

4

The distinction has, however, not been wholly accepted by American authorities. Dr. DA COSTA questions the main point of difference —the temperature; and Dr. ALFRED STILLÉ writes: " The whole medical literature does not contain a single case of *sporadic* idiopathic cerebro-spinal meningitis with the characteristic *sudden onset* of the epidemic disease." From that writer's admirable monograph and his article in Pepper's System of Medicine, we extract the following exhaustive comparison of meningitis and typhus,* with which it has often been confounded:

EPIDEMIC MENINGITIS.	TYPHUS FEVER.
A pandemic disease. Occurs simultaneously in places remote from one another, and without intercommunication.	An epidemic disease, due to local causes and spreading by intercommunication.
Attacks all classes of society. Is never primarily developed by destitution, squalor, or defective ventilation.	Attacks the poor, filthy, and crowded alone.
Is not contagious.	Contagious in a high degree.
Attacks more males than females.	Both sexes equally affected.
Attacks more young persons than adults.	More adults than young persons.
Generally occurs in winter.	Epidemics irrespective of season.
Eruptions are absent in at least half of the cases; they occur within the first day or two.	Eruption rarely absent, and appears about the fifth day.
The eruptions are various; they include erythema, roseola, urticaria, herpes, etc. Ecchymoses are common.	Eruption always roseolous and then petechial. Ecchymoses are rare.
Headache is acute, agonizing, tensive.	Headache dull and heavy.

* Stillé. Pepper's System of Medicine, Vol. I, Philadelphia, 1885, page 827.

EPIDEMIC MENINGITIS.

Delirium often absent; often hysterical, sometimes vivacious, sometimes maniacal. Generally begins on the first or second day.

Pulse very often not above the natural rate; often preternaturally frequent or infrequent. Is subject to sudden and great variations.

The temperature is lower than that recorded in any other typhoid or inflammatory disease. It is also very fluctuating.

The body has no peculiar smell.

The tongue is generally moist and soft, and if dry, is not foul. Sordes on teeth rare.

Vomiting is an almost constant and urgent symptom, especially in the first stage.

Pains in the spine and limbs, of a sharp and lancinating character, are usual.

Tetanic spasms occur in a large proportion of cases, and within the first two or three days. They are due to an exudation on the medulla oblongata and spinalis.

Cutaneous hyperæsthesia is a prominent symptom.

Strabismus is common.

The eyes, if injected, have a light red or pinkish color.

The pupils are often variable and unequal.

TYPHUS FEVER.

Delirium rarely absent; usually muttering. Rarely begins before the end of the first week.

A slow pulse exceedingly rare. Its rate usually between 90 and 120.

The temperature is always elevated, and does not fall until the close of the attack. The skin is hot, burning, and pungent to the feel.

The mouse-like smell is characteristic.

The tongue is generally dry, hard, and brown, and the teeth and gums fuliginous.

Vomiting is rare and not urgent.

The pains, if any, are dull and apparently muscular.

Tetanic spasms are unknown in typhus. Convulsions sometimes occur, due to pyæmia.

The sensibility of the skin is usually blunted.

Strabismus is rare.

The blood in the conjunctival vessels is dark.

The pupils are equal and contracted.

EPIDEMIC MENINGITIS.	TYPHUS FEVER.
Deafness and blindness are often complete and permanent.	Deafness almost always ceases with convalescence. Blindness never follows typhus.
Duration very indefinite, but generally from 4 to 7 days.	Duration from 12 to 14 days.
Relapses are common.	Relapses are rare.
The blood is often fibrinous.	The blood is never fibrinous.
The lesions, except in the most rapid cases, consist of a plastic or purulent exudation in the meshes of the cerebro-spinal pia mater.	In typhus no inflammatory lesions exist.
Mortality from 20 to 75 per cent.	Mortality from 8 to 40 per cent

ACUTE TUBERCULAR (GRANULAR) MENINGITIS.

This serious disease, which usually occurs during adolescence, is apt to be confounded, especially in the adult, with typhoid or typhus fever, the exanthemata, and pneumonia.   The following characteristics of the disease, as given by Drs. REGINALD SOUTHEY and HAMILTON, will serve to distinguish it:

1. The prodromal symptoms of this form of meningitis are well marked.   The history of the case usually records an illness that has endured some two or four weeks, but one which has not attracted much attention until distracting headache, with some delirium at night, has supervened.

2. Vomiting is generally the first and most important symptom. Headache is invariably present.

3. After two or three days there is a marked rise of temperature, say from 101° to 105°, with greatly increased pulse.

4. The bowels are constipated and not tender to firm pressure. Very little nourishment is voluntarily taken.   The abdomen becomes retracted, and the aspect of the patient, with half-open eyelids, or some slight paralysis of these, becomes highly diagnostic.

5. There is no characteristic rash.   The so-called *tâche cérébrale* of this form of meningitis is not a true eruption, but is produced by pres-

sure or contact. When the finger is drawn across the skin of the forehead it leaves a vivid red mark, which has been considered a pathognomonic sign of the disease.

6. The skin is hyperæsthetic, the delirium slight and transitory, the
temper irritable, obstlnate and unaccommodating.

7. There are general muscular pains, followed first by stiffness, and
then by slight paralysis, as shown in the imperfect coördination of the
muscular movements also in tremblings and twitchings. The muscular
pain and stiffness are often first complained of in the nape of the neck,
and then in the muscles of the back.

8. Slight epileptiform convulsions are observed, followed by paralysis of motion in the limbs or parts convulsed; this paralysis being
most usually of a transitory or temporary kind. Among the paralyses most characteristic are those affecting the optic commissure and
oculo-motor tracts, causing a slight internal squint, with dilated in-
active pupil of one eye, with drooping of the same eyelid, and paralysis of the facial nerve upon one side. The paralysis of the limbs,
although usually hemiplegic, is seldom one that invades the body
upon one side in its entirety. Further, its mode of attack is gradual;
usually, the arm and leg are affected upon the same side, even when
the facial muscles are not involved.

YELLOW FEVER.

The name *Yellow Fever* is misleading, as the coloration of the skin
to which it refers is not an invariable nor even a common sign of the
disease. According to Dr. GREENSVILLE DOWELL,* the skin does not
turn yellow in more than one case in six, and many die before there
is the least appearance of yellowness even in the eyes. Of those who
die after the black vomit has set in, not more than one in three presents the yellowness.

The most pathognomonic sign of the disease is the *black vomit.* It
is brownish-black, semi-fluid, with a glistening reflection and acid reaction, and varies in quantity from a mere stain on a handkerchief to
many pints in the twenty-four hours. It, however, is not thrown up
in more than one in three fatal cases.

* " Yellow Fever and Malarial Diseases."

The usual course of the disease as witnessed in the southern and southwestern States is as follows:

1. Onset with a chilly feeling along the spine, passing into actual rigor.

2. Pain in the head, severe in proportion to the malignancy of the disease.

3. Fever slight, tending to perspiration.

4. Remission after a period varying from twenty-four hours to five days.

5. The secondary fever, commencing usually without a chill; it runs an indefinite course.

The discoloration begins at the white of the eye, and extends over the skin of the forehead, chest, abdomen, and extremities. The urine is high-colored, and stains linen, and in some cases the perspiration gives the same yellowish stain.

The shades which separate the symptoms of one fever from those of another, in warm climates, are sometimes of such gentle gradation that *primâ facie* they may seem to belong to one and the same disease; this more especially refers to the yellow and remittent type of fevers, between which so slight is sometimes the distinction, that bilious remittent has frequently been considered and classified as true yellow fever; for in the prominent symptoms which appear in both yellow and remittent fever, a great similarity obtains: both take their origin in paludal soils; both in their course offer symptoms of so seemingly similar a nature that the shades of difference are so slight as to frequently escape even a good observer, and cause him to fall into error. But this apparent similarity vanishes on close and continuous inspection, for then essential and distinctive marks are observed, which stamp each with an individuality, and which characterize each as a separate disease, distinct in its essence, and differing signally the one from the other. These differences may be summarized as follows (J. J. L. DONNET, DA COSTA, DOWELL, and others):

YELLOW FEVER.	BILIOUS REMITTENT.
Is essentially of an infectious nature, and found in cities.	Is not of an infectious nature, and usually found in the country.

YELLOW FEVER.

Chiefly vigorous and young constitutions fall victims to it. Colored population *less* liable than white.

Restricted chiefly to the yellow fever zone.

Is of a continued type; remissions not marked.

Temperature in bad cases very high.

Usually attacks at night.

Severe nausea and vomiting throughout. Epigastric tenderness early and decided black vomit. Headache occipital.

Hemorrhages from the gums and various parts of the body.

Tongue clean or but slightly coated; pulse variable, becoming slow in the last stages.

Eye highly injected and humid; expression often fierce or anxious.

Pain in the back very severe; also pain in the calves and over the eyes.

Delirium rare; mind generally clear and cheerful.

Urine generally albuminous; suppression common.

Muscular prostration slight; convalescence rapid; no sequelæ.

Liver affected and slightly enlarged.

Spleen not affected.

One attack affords an almost certain immunity.

BILIOUS REMITTENT.

All ages and constitutions are liable, and the weakest most so. Colored population liable.

Is to be found in all parts of the world where marshy soils prevail.

Remissions observed in the morning.

Temperature not extraordinarily high.

Usually attacks in daytime.

Nausea and vomiting moderate. Epigastric tenderness slight. Headache frontal.

No hemorrhagic tendency. .

Tongue heavily coated; pulse varies little, remaining quick until convalescence sets in.

Eye and physiognomy not peculiar.

Rachialgia slight or absent; headache moderate.

Delirium frequent; mind always dull.

Albuminous urine rare; suppression also rare.

Much muscular prostration; convalescence slow; sequelæ various and tedious.

Liver not affected.

Spleen invariably affected.

One attack seems rather to predispose to others.

YELLOW FEVER.	BILIOUS REMITTENT.
Mortality very high.	Mortality slight.
Peculiar smell often perceptible.	No peculiar smell observed.
Never merges into intermittent.	Often merges into intermittent.
Treatment unsatisfactory; quinine useless.	Quite amenable to treatment; antagonistic power of quinine beyond question.
Autopsies show great congestion, inflammation, ulceration, and softening of the stomach. Liver enlarged, fatty, yellowish in color, its secreting cells filled with oil globules. Heart often exhibits disintegration of the muscular fibres.	Autopsies show congestion of the stomach, but rarely inflammation. Liver of an olive or bronze hue, not fatty. Spleen enlarged.
Micrococcus xanthogenicus (?) in blood and the urine (Freire).	Plasmodium malariæ, or malarial bodies of Laveran, can be found in the blood.

RELAPSING FEVER.

Of late years epidemics of this disease have appeared at various points in this country. It is eminently contagious in character, and a physician should be prepared to recognize it early. The invasion is sudden, the fever soon developed and high, the pulse very rapid, the skin often jaundiced, and the temperature elevated (106°–107°). Toward the close of the first week the symptoms rapidly subside, and convalescence seems at hand; but after about another week the symptoms all return with as much violence as ever, to again disappear, as a rule, after four or five days.

The epidemic prevalence of the disease, its sudden invasion, the persistence without remission of the high febrile symptoms, and the afebrile interval, give it a peculiar physiognomy.

The characteristic feature of the disease, asserted by some to be truly pathognomonic, is the presence in the blood of the *spirillum Obermeyeri*. The following method of demonstrating them is recommended by Dr. R. ALBRECHT, of St. Petersburg: *

* St. Petersburg *Med. Wochenschrift*, June, 1878.

Spread out a drop of blood on a slide, not too thin; let it dry; treat it with a drop of acetic acid, and repeat it in a few seconds. By this means all the fibrin and blood-corpuscles will be destroyed and dissolved, and after careful washing away of the acid with distilled water, and final drying, the preparation is ready for use. With a little care in washing, which must not be in a stream, the spirilla are not lost, especially if the preparation has been dried for six to twelve hours before being treated with acetic acid. The glass slide then looks quite transparent, and, at the place where the drop of blood was, it looks a little dusty. Under the microscope the nuclei and nucleoli of the white blood-corpuscles are visible, and between these the spirilla lie in great numbers and in the most distinct arrangement and position, showing up very beautifully and distinctly. They give the impression of being thicker than they generally are, probably because they are no longer imbedded in a highly refracting substance—plasma.

Relapsing fever is liable to be mistaken for one of the forms of continued fever. Its epidemic prevalence will naturally put the physician on his guard. It is, moreover, especially a disease of the lower classes, who suffer from insufficient food and filthy surroundings. In most cases it is associated with jaundice, which is a rare complication in typhoid. When the disease rapidly abates, and this·cessation is followed by the characteristic relapse, no reasonable doubt as to its nature can be entertained. The main distinctions between relapsing and typhoid may be thrown into a comparative view, as follows:

RELAPSING FEVER.	TYPHOID FEVER.
Invasion sudden.	Invasion gradual, with epistaxis; no chill.
Bowels generally constipated.	Generally diarrhœa.
Conjunctivis occurs early.	No conjunctivitis; eyes bright and clear.
Liver engorged, skin yellow, tenderness over epigastrium.	No yellowness; tenderness over right iliac region.
Temperature high, 105°–107°.	Temperature during first week rarely above 104°.

RELAPSING FEVER.	TYPHOID FEVER.
Fever abates in three or four days, with critical sweats; diminution or cessation of the febrile symptoms, with subsequent relapse.	These phenomena absent; symptoms continuous for three or four weeks.
Spirilla in the blood.	No spirilla.
Splenic enlargement.	Spleen only moderately enlarged.
No characteristic eruption.	"Rose spots," inflammation of Peyer's glands.

CHAPTER II.

DISEASES OF THE BLOOD.

THE DYSCRASIÆ.

As is justly remarked by Professor THEODOR BILLROTH, in his *Surgical Pathology*, while it is true that there are some objections to the employment of the term *dyscrasia*, as committing one to a humoral pathology, these are overbalanced by the fact that there are certain well-defined, long-recognized, inherited physical peculiarities, which render the person possessing them unusually prone to certain diseases and complications, and which lend a complexion of their own to very many affections seemingly remote in form and pathology.

These constitutional tendencies may as well be known by the term *Dyscrasiæ* as by any other, since their existence cannot well be denied.

The principal dyscrasiæ are: 1. The *arthritic*, sometimes called dartrous or rheumatic, believed to be pathologically akin to lithæmia, gout, and rheumatism; 2. The *strumous*, or scrofulous; 3. The *tuberculous*, or phthisical; the last two mentioned, in the opinion of some, being merely the outgrowth of an inherited syphilitic taint.

I. THE ARTHRITIC, DARTROUS, OR RHEUMATIC DYSCRASIA.

This form of blood poisoning has been aptly termed, by Mr. JONATHAN HUTCHINSON, "the basis-diathesis on which both gout and rheumatic arthritis are built, and which, to a large extent, is indifferent and common to both." When a man with such a diathesis becomes affected with a renal disease, gout develops itself; otherwise he will probably have rheumatism. In many families it has been observed that

(59)

the males have gout, the females rheumatism. The explanation is not far to seek.* In another lecture Mr. Hutchinson describes gout as " chronic rheumatism *plus* a dietetic derangement."

Many skin diseases, nervous affections (so-called), "cramp colic," headaches, sciatica, vertigoes, palpitation, and obstinate dyspepsia, are really latent gout (lithæmia). In such cases there is usually a history of antecedent or hereditary rheumatic diathesis, frequent acid eructations, the emission of pale, limpid, acid urine, of low specific gravity, and with traces of sugar or albumen, or both; some varicosity of the veins; the nails are brittle; and there is slight redness around the eye, indicative of mild chronic conjunctivitis (Dr. J. RUSSELL REYNOLDS).

The following are the signs, as stated by Professor HARDY, of Paris: Persons who have this diathesis appear to enjoy good health, but their skin is habitually dry and their perspiration scanty. They often experience a lively itching without eruption. The appetite is generally well developed, and they are apt to eat a much greater quantity of food (especially animal food) than others in analogous conditions. Another important peculiarity is the extreme sensibility of the skin, and the facility with which it is influenced by the lightest and most fugitive impressions. Sometimes general excitement, alcoholic excess, watching, use of coffee, of certain kinds of food; sometimes a local excitement, irritating frictions, or the application of a plaster, will give rise to an eruption, often ephemeral, which reveals a peculiar predisposition of the economy, and the existence of a latent vice which needs but a favorable occasion to manifest itself.

To this diathesis HARDY ascribes eczema, lichen, psoriasis, and pityriasis, among diseases of the skin.†

Mr. PRESCOTT HEWETT adds that when a patient complains of dyspepsia, more or less troublesome, frequent deposits of lithates in the urine, slight eczematous eruptions on the skin from time to time, anomalous wandering pains in various muscles, sharp, deep-seated pains in the tongue, continuing for or two three days, and then disappearing altogether for a while, crackling about the cervical spine on

---

* *Medical Times and Gazette*, June, 1876.

† *Maladies de la Peau*, Paris, 1860.

slight movements, some, it may be very slight, knottiness about the smaller joints of the fingers—we may be very certain that he has the arthritic diathesis.

Sir JAMES PAGET adds to the above: Small (chalky) nodules in the cartilages of the ears (*tophi*); nodular enlargement of the knuckles; thickening of the cutis, with subcutaneous bursæ over the knuckles, chiefly between the first and second phalanges of the fingers; thickening of the palmar fascia, adhering to the cutis, and producing contraction of the fingers; spontaneous pain in the tendo-Achillis; pain in the heel; frequent and persistent erections at night, not connected with any sexual feelings; "burning soles" and "burning palms;" sensations of heat; tingling and burning patches of the skin of the thighs, without external appearances of redness or eruption; patches of "dry eczema."

In such patients an injury may be followed by a well-marked attack of gout; or the trouble may linger, with pain and occasional swelling, and with constantly increasing distrust of surgery and the surgeon, till some one suspects the existing taint of the arthritic diathesis, and acting on the suspicion, addresses his remedies to it, and promptly cures the local trouble. The disorder known as rheumatic fever, or acute articular rheumatism, has been ascribed by MACLAGAN to an infection introduced from without into a system predisposed to the influence. Heuter, Reckinghausen, and Klebs declare the active agent to be a variety of bacteria—a micrococcus.

(The points of diagnosis between the gouty diathesis and chronic rheumatism, as summed up by FOTHERGILL, are given on page 65, in the section devoted to "Diseases Likely to be Confounded with Rheumatism.")

II. THE SCROFULOUS OR STRUMOUS DYSCRASIA.

Sir JAMES PAGET defines the principal signs of scrofulous constitution to be the occurrence of slowly progressive and long abiding inflammation, provoked by less causes than would excite inflammation in healthy persons, the inflammatory process tending to the production of "cheesy" matter; the middle permanent incisors, with their borders barred, crenated, thin and brittle; the mucous membrane of the lower

turbinated bone swollen, puffed and congested; a long abiding ozæna
in early life, with frequent or daily discharge of scabs; general swelling, with glandular enlargement of the whole naso-palatine mucous
membrane; a granular pharanx, with its lining membrane more or less
thickly scattered with prominent glands; the perforating ulcer of the
nasal septum—these are some of the minor signs. Still more positive
are enlarged and suppurating lymph-glands discharging curdy pus,
and slowly healing with red-banded and barred scars; pustules by the
edge of the cornea; frequent impetigo with swollen glands; periosteal
swellings of the phalanges; chronic thickenings of synovial membranes; obstinate otorrhœa. If a patient is found to have or to have
had any few of these, he may justly be pronounced scrofulous, and
scrofula may be suspected in any localized morbid process in him.
Or, if these diseases are known to have occurred singly or together in
many members of a family, we should look out for scrofula as an element of whatever disease may appear in any member of that family.

Dr. FRANCIS DELAFIELD, of New York city, observes* that practitioners in this country see so little of scrofula comparatively that it is
difficult for them to appreciate the prominent place it holds in the
minds of physicians in European countries. It is a condition which is
hardly susceptible of a definition, and yet it is not hard to understand
what is meant by the term.

It means this: When an individual acquires an inflammation of the
mucous membrane, of the skin, of the joints, of the bones, of the
genito-urinary apparatus, or of almost any part of the body, such an
inflammation usually runs an acute course and terminates in resolution,
or in suppuration, or in the formation of organized new tissue. But,
if the inflammation, instead of doing this, simply reaches a certain
point and stays there, and then, instead of resolving or suppurating
merely, goes through a succession of degenerative changes, such an
inflammation is said to be scrofulous.

The scrofulous inflammations have several well-marked characteristics. They are very slow in their progress; they are very rebellious
to treatment; they are accompanied by an extensive cellular infiltration of the inflamed parts, so that when the degenerative changes en-

sue there is large destruction of tissue. The degeneration which occurs in the products of such a scrofulous inflammation is peculiar in its nature; it is commonly called cheesy degeneration, and consists in the transformation of the products of inflammation into a dry, yellow mass, composed of amorphous granular matter. Examples of this form of inflammation will at once suggest themselves. Caries of the vertebra, hip-joint disease, white swelling of the knee-joint, scrofulous orchitis, and enlarged lymphatic glands, are all of frequent occurrence. In a number of instances of scrofulous inflammations, an examination of the caseating product has revealed the presence of bacilli, which present characters almost if not quite identical with those of tuberculosis bacilli[*] to be discussed on another page. (See page 65.)

III. THE TUBERCULOUS DYSCRASIA.

There are families in which the children, while apparently healthy during their development, perish early in adult life with tubercular manifestations, especially in the lungs. This indicates a peculiar inheritance, which may be called the tuberculous dyscrasia. More frequently the children of decidedly strumous patients die in infancy, with tubercular meningitis, which furnishes ground for the belief that in many instances tubercular disease is brought about by the strumous dyscrasia; and, indeed, it is by many identified with it. The physical characteristics of scrofulous subjects belong also to the majority of consumptives, in a greater or less degree. Others are predisposed to the disease through defective oxygenation caused by unfavorable form of the thoracic walls. But the researches on this subject are still incomplete, and it is well to bear in mind the words of Dr. A. T. H. WATERS:—

"There is no temperament which does not furnish victims to consumption; nor can we say that there is any conformation of the body which is characteristic of the phthisical. I have seen men and women with the best developed frames and the most ample chests attacked with phthisis. You must not, therefore, be misled, by the existence of these conditions, by the appearance of robustness in your patients, into imagining that they cannot possibly become the subjects of this disease."

[*] Klein. Micro-Organisms and Disease. London, 1884, p. 118.

The diagnosis of these different conditions is made less difficult by bearing in mind their peculiar tendencies and characteristic manifestations, as set forth in the followiug table:

SCROFULOSIS.	TUBERCULOSIS.	INHERITED SYPHILIS.
More particularly limited to childhood.	Not specially limited.	Manifests itself early, generally before third month (from fourteen days to six weeks).
Affects especially the lymphatic glands (causing abscess), the mucous membranes (ophthalmia), the skin (obstinate cutaneous diseases, especially the pustular) and bones (caries and necrosis,) indolent abscesses (cold abscess). Frequently resulting in phthisis and hydrocephalus.	Affects internal organs (phthisis, hydrocephalus, peritonitis, tabes of mesenteric or bronchial glands).	Prominent symptoms:— Cachectic appearance, snuffles, condylomata around the anus. Child thin, poorly nourished, muscles flabby. Skin brownish, cracked, thick and rough. Fontanelle open; ossification slow. Posterior cervical glands enlarged. Second set of incisors characteristic (Hutchinson teeth). The central incisors short, narrow and thin, chisel-shaped; edges soon become notched and broken; also striped or ribbed horizontally. Hair thin, and may have alopecia. Eruptions copper-colored and chronic; generally dry, but may be pustular (erythema, lichen, psoriasis and eczema, or impetigo, ecthyma and pemphigus), often seen on palms of hands or soles of feet. Liver enlarged (albuminoid). Ascites common, with tympanites.
Generally afebrile.	Pyrexia marked in acute cases, irregular in chronic.	Pyrexia only in complications.
Temperament phlegmatic; mind and body backward; skin muddy; upper lip thick; nostrils wide and alæ thickened. Abdomen tumid; ends of bones large; shafts thick. Otorrhœa, ozæna, ophthalmia common.	Nervous system highly developed; mind and body active; organization delicate and refined. "Thin skin, clear complexion, the surface veins distinct, eyes bright, pupils large, eyelashes long, hair silken, face oval, ends of bones small, shafts thin, limbs straight." (JENNER).	Physically and mentally inferior in structure, and slow in movement; inactive, dull and often cachectic looking.
Bacilli present in caseating glands.	Bacilli in products.	No bacilli detected.
Mercury prohibited.	Mercury only for temporary indigestions, etc., used carefully.	Mercury the sheet-anchor.

They resemble each other in hereditary character, their familiar manifestations being readily induced by defective hygienic conditions.

EHRLICH'S METHOD FOR DETECTING TUBERCLE BACILLI.

A power of 400–500 diameters is required of the microscope which should also be provided with a sub-stage condenser to flood the field with light. "A fragment of thick opaque sputum is to be taken in forceps, placed on a cover glass, and spread in a thin layer by means of a second cover glass. The prepared slide is to be passed slowly through an alcoholic flame or that of a Bunsen burner, till the layer of sputum is dried. A saturated alcoholic solution of methyl-violet, or fuchsin, is made and filtered and added, drop by drop, to a filtered saturated solution of aniline oil, shaken in water. The color is to be added with stirring till an opalescent film forms on the surface of the mixture. The slide containing the dried sputum is to be placed in or on this staining fluid, and allowed to remain for half an hour or less, the application of warmth hastening the process, when it is removed and the specimen is decolorized in a solution of one part of nitric acid in two parts of water. The preparation is then washed in water and may be examined directly in water, glycerin, or, after dehydration in alcohol, in oil of cloves. The tinted bacilli are made more prominent by a secondary staining, for a minute or two, of the red (fuchsin) preparation in a concentrated solution of methyl-blue, the violet preparation being secondarily stained in a like solution of aniline brown. If the preparation is to be permanently preserved, it should be dehydrated in strong alcohol after washing in water, and it may then be treated with oil of cloves and mounted in Canada balsam.*

DISEASES LIKELY TO BE CONFOUNDED WITH RHEUMATISM.

Ordinarily an attack of acute rheumatism is recognized without difficulty by the pains in the joints, their swelling and tenderness, the shifting character of the disorder from joint to joint, and the absence of the symptoms so common in continued fevers, of disturbance of the

* Pepper's System of Medicine. Phila., 1885. Vol. I, page 102.

5

stomach and brain (if we except the so-called cerebral rheumatism which appears to be associated with a urǣmic condition, if its symptoms are not in reality occasioned in this way), as well as of the intermissions or remissions of periodic fevers.

Nevertheless it is true, as remarked by Dr. S. O. HABERSHON,* that while there are many characteristics of true rheumatic diseases, few maladies are more easily mistaken, and there is *no* sign which is *uniformly* present. Pain is, perhaps, the most constant indication, with stiffness of one or more joints; but rheumatic pericarditis may, and often does, exist without any pain whatever. The same may be said in reference to febrile symptoms, to increase of temperature, and to changes in the urine; none of these signs is pathognomonic. Many maladies are designated rheumatic which have no connection with that disease.

1. *Diseases of the spine* are often said to *commence* with an attack of rheumatism; but it will generally be found that the pain in the course of the nerves or in the fibrous tissues arises from direct implication of the nerves or their centres.

2. The same remark applies to pain produced by the pressure of *cancerous*, *aneurismal*, or other tumors. Thus cancerous disease of the lumbar glands is often mistaken for lumbago; so also the pain from aneurismal disease of the thoracic and abdominal aorta, early in the disease, when no pulsating tumor can be detected, is generally referred to rheumatism.

3. During the course of *renal disease*, abnormal irritation arises not only in the serous membranes, producing pericarditis, pleurisy, peritonitis, etc., but a similar change happens with the synovial membranes, and a form of disease is induced which simulates rheumatism.

4. In chronic poisoning by *lead*, vague pains in the fasciæ, as well as in the joints, have been designated "saturnine arthralgia."

5. *Periosteal disease* is occasionally a source of fallacy in the diagnosis of rheumatism.

6. *Shingles* or *herpes zoster* may be found in the course both of the cerebral and spinal nerves; and the severe pain which precedes the

---

* *Half-Yearly Compendium of Medical Science.* III.

eruption of the vesicles, and which also follows their disappearance, closely simulates intercostal rheumatism.

7. A more important disease, and one which is attended with fatal issue, is *pyæmia*. It closely resembles rheumatism; for, with rigor and febrile symptoms, there is fixed pain and swelling in the joints— first one and then another being affected, *though without subsidence of those parts first attacked*. But while there may be some similarity in the symptoms, the prognosis is widely different. The one is generally a curable disease, the other a fatal one.

8. *Acute synovitis* closely resembles rheumatism, having pain and heat in the joint, with distention. But as a rule it affects only one joint; it is never subject to metastasis; and there is little or no effusion into the surrounding tissue. The accumulation of fluid in the joint is greater, but the constitutional symptoms are less prominent.

9. *Milk leg* occurs after fevers, or, in women after confinement. The limb swells throughout, becoming white, firm, hot and shining, and pits but little on pressure. The history of the case and appearance of the limb are usually sufficient to form the diagnosis.

10. *Myalgia* (myodynia) has no essential relation to rheumatism or the rheumatic diathesis. The common name of muscular rheumatism is incorrect, for rheumatism never primarily or exclusively manifests itself in inflammation of muscle substance. It is a disease of nutrition, and not a diathetic disease,* but sometimes shows decided hereditary tendency. Its varieties are cephalodynia, torticollis, pleurodynia, lumbago, dorsodynia, omodynia, scapulodynia, etc. Growing pains of childhood are of this category, and certain pains of the praecordia in cases of degeneration and lesions of the heart are without doubt myalgic in character. Pain, tenderness, and in chronic cases, spastic rigidity localized in certain muscles or groups of muscles without fever, or other obvious disturbances of the general health, are the prominent characters of myalgia. (Neuralgia and myalgia are contrasted later in this work).

CHRONIC RHEUMATISM.

The most common form of chronic rheumatism is that which affects the muscles, and it is frequently by no means easy to distinguish the

* J. C. Wilson, Myalgia, *Phila. Med. Times.* Vol. xvi. p. 120.

pains due to the rheumatic diathesis from those of a wholly diverse etiology.

The principal distinctions are:

1. From *neuralgia*. Neuralgic pains are usually confined to the dis tribution of one nerve ; they are not increased by motion or pressure; they are not attended with diffused soreness; and they are variable in intensity, and are not attended with acid secretion. The tender spots of Valleix may be detected along the trunk of the nerve or in its distribution.

2. *From the pains of organic lesions.* These are usually so clearly localized as to point to their origin. Nevertheless the pain in the right shoulder, symptomatic of hepatic disease, and especially of an abscess approaching the serous surface of the liver, and the sympathetic pain down the left arm in some cases of heart disease (angina pectoris), are often carelessly overlooked, and their significance unheeded, by classing them as rheumatic. Intercostal rheumatism has included pleurisy, pleurodynia, broken ribs, herpes, neuralgia, the peculiar pain, generally of the left side, found in women and connected with menorrhagia and leucorrhœa; the pain on either side, which is intimately connected with debility and anæmia; and again it is very often confounded with that condition of pain and soreness of the muscle developed by overwork, and attended with both muscular aud cutaneous hyperæsthesia, designated by INMAN "myalgia."

3. *From the osteocopic pains of syphilis.* The history of the case throws some light; but as this often cannot be had, it should be remembered that syphilitic periostitis evinces a decided partiality for the periosteum and shafts of the long bones, and is very generally accompanied by nodes, especially in the anterior surface of the tibia, which are almost pathognomonic. There is often, too, a more marked cachexia than is found along with non-specific rheumatism. The clavicle, humerus, and forearms, are frequent locations of this form of rheumatism. As well as its favorite seats and accompanying nodes, there are evidences of skin and throat affections, a mutilated iris, etc., which will assist in forming a correct diagnosis. Furthermore, the ready response to a specific treatment aids in distinguishing syphilitic pains.

4. *From progressive locomotor ataxia.* Ataxic patients often bitterly

complain of supposed rheumatic pains. These pains, in locomotor ataxy, come on in severe pangs—"stabbing, boring, shooting like lightning, flitting from one place to another in a very erratic manner, and recurring in paroxysms lasting from a few minutes to many hours." Their suddenness is their especial characteristic, and should always put the medical observer on his guard to look out for the other indications, as loss of tactile sensibility, etc. These pains may be accompanied by a feeling of coldness, thus closely simulating some forms of rheumatism. The importance of them lies in the prognosis, as the ' causes of locomotor ataxy are not to be relieved by art, although the pains may be mitigated by anodynes and frictions.

5. The *pains of chronic renal disease* often closely simulate *lumbago*, or muscular rheumatism of the loins. No clinical distinction can be positively drawn, except from examination of the urine; but, in some forms of renal disease albumen is often absent for long periods together. Moreover, the amount of the urine varies, and when great in quantity is usually of low specific gravity, and contains granular casts, which, however, are often few in number and not easily found. An absolute diagnosis is not always attainable. In gouty kidneys we may fall back upon the rational symptoms, and the distinguishing characteristics may be found to run in the following directions : Rheumatism is associated with the fibro-serous texture ; in lithæmia the poison has more affinity for the true serous surfaces, and is often the cause of pleurisy and peritonitis. Lithæmia more affects the muscles, and rheumatism rather the large joints. Diarrhœa, vomiting and other affections of the mucous membrane, as bronchitis, accompany lithæmia; and in these it differs from rheumatism. Lithæmia is accompanied by headache, especially of the vertex (persistent and recurring vertical headache is almost pathognomonic of lithæmia), or the pain may be frontal. (FOTHERGILL.)

6. A dislocation of the shoulder has been prescribed for as "rheumatism," which shows the necessity of inspection of affected joints.

A typical effect of the acid diathesis of chronic rheumatism is the *rheumatic markings of the teeth*, to which attention has been directed by Dr. L. G. NOEL.*

* Nashville *Journal of Medicine and Surgery*, Feb., 1875.

These markings seldom appear until after middle life is past. They are most frequent upon the crowns of the teeth, though they are sometimes seen upon their buccal and labial surfaces. It is that condition of the teeth treated of in dental works as "spontaneous abrasion."

The abrasion often begins as decay in the fissures on the grinding surface of the molars and bicuspids, but instead of following the tubuli, and dipping deep into the interior of the teeth, these become closed by a calcareous deposit, as fast as laid open, and the decay spreads out into a wide saucer-shape. This cupping out of the teeth is not, however, confined to the molars and bicuspids, but commencing upon the cusps of the canines, and cutting edges of the incisors, as mere mechanical abrasion, asperities disappear, the teeth become square and polished on the ends, and presently the surfaces begin to assume a concave, instead of their original convex, appearance. This cupping out may go on until the pulp is so nearly reached as to become irritated to the point of inflammation and death; but usually its irritation is only sufficient to cause a deposition of secondary dentine on the interior of its chamber, a part of its substance forming a matrix, in which lime-salts are deposited.

Syphilitic lesions of the teeth are best seen in the upper central incisors of the second dentition (Hutchinson teeth) chisel-shaped with cupped free border.

Osseous lesions of inherited syphilis may be distinguished from those of Rickets by attention to the following points :*

INHERITED SYPHILIS.	RICKETS.
The swellings, particularly those of the long bones, show themselves at or soon after birth.	Rarely appear before six months, generally still later.
A history of syphilis or evidence* of existing syphilis in one or both parents.	No such history necessarily.
Preceded or accompanied by snuffles, coryza, and cutaneous and mucous lesions.	No such prodromata.

* J. Wm. White. Hereditary Syphilis. Pepper's System of Medicine, Vol. ii., p. 290.

INHERITED SYPHILIS.	RICKETS.
No such prodromata in most cases.	Pallor, restlessness, sweating, nausea, diarrhœa, etc., constitute a combination of symptoms which often precede the bone disease.
Cachexia absent or moderate.	Cachexia marked.
Physiognomical peculiarities of syphilis present.	Not present as a group.
Circumscribed tumors on frontal and parietal bones, rarely on occiput.	Cranial bones thickened in spots, usually upon occiput.
Ribs not markedly affected.	All, or nearly all involved.
Swellings on long bones or extremities irregular.	Extremities symmetrically enlarged.
Disease of ribs, when existent, not ordinarily coincident with that of other bones.	Nearly always so.
Fontanelles close at an early period.	Closure delayed.
Other syphilitic symptoms present, enlargement of phalanges, metatarsal bones, etc.	Syphilitic symptoms absent.
Often accompanied by sinuses, synovitis, abscesses, cutaneous ulcers, etc.	Little external or surrounding involvement.
Generally disappears by resolution, without leaving any permanent change.	Usually leaves some bending of shaft and distortion of the neighboring joint.
Mortality among children in whom many bones are involved is very great.	Much less.
Specific treatment useful.	Of no benefit.
In the first stage there is an exuberant calcification of the ossefying cartilage causing necrosis of the new formed tissue and a consecutive inflammation, which terminates in the separation of the epiphyses.	This is less marked. There is formed, instead, a soft and non-calcified osteoid tissue.

GOUT.

The signs of gout have already been in part referred to (page 60). It is not nearly so frequent in the United States as in England, and is apt, therefore, to be mistaken for rheumatism, which it closely resembles.

The following table of differences will facilitate the diagnosis:—

GOUT.	RHEUMATISM.
Generally hereditary.	Rarely hereditary.
Occurs usually in men, beyond middle age, rarely in women.	Occurs oftener in women, and before middle age.
Attacks generally periodic, and last about a week.	Attacks dependent on exposure, and last several weeks.
The small joints chiefly affected, especially that of the great toe, or lower extremity.	The large joints are those generally attacked.
Much local pain, redness, œdema, and enlargement of veins.	All these symptoms less marked.
Kidneys generally affected; little fever; no sweating; heart not implicated.	Kidneys not involved; fever often high; sweating profuse; heart often implicated.
Chalk stones in the joints and ears.	Chalk stones never present.
Uric acid always present in the blood in large excess (GARROD).	Uric acid never found in excess. Increase of fibrin in the blood.*

Dr. GARROD says that the presence of uric acid in the blood can readily be demonstrated by taking a fluidrachm of the serum from a blister, adding to it six minims of acetic acid, and placing a thread in the mixture. The uric acid, if present, will be deposited in fine crystals along the thread.

---

* Hayem. The increase of fibrin in the blood may precede any local manifestation of the disease; so that by examination of the blood at the onset of the fever one may diagnosticate the rheumatism in advance and thus prevent joint complications.—*Phila. Med. Times*, Vol. xvi., p. 504.

RHEUMATOID ARTHRITIS (RHEUMATIC GOUT, ARTHRITIS RHEUMATICA DEFORMANS).

This is by no means an infrequent disease in this country, and is a very serious one. It is now acknowledged by the best authorities to be a distinct malady, different in origin, history and treatment from both rheumatism and gout. It is common in women and young persons, and is not produced by alcoholic or other excesses. It implicates joints of all sizes, and in all the extremities. They become permanently affected, stiffened and enlarged, but no deposits of urate of soda are found in them. The disease frequently shows itself without fever; the joints swell by serous effusions into the capsules, and along with this the ends of the bones enlarge. The integument is not inflamed, or but moderately so, and the muscles do not appear to suffer. The result on the joint may be subluxation, relaxation, or anchylosis. The concretions attendant on the disease prove, on analysis, to be of the same composition as bone, with a slight preponderance of lime (DRACHMANN). Phosphoric acid is diminished in the urine and increased in the blood (BöCHER).

Neither the treatment of gout nor that for acute rheumatism yields its usual results in this disease.

PERNICIOUS ANÆMIA AND LEUKÆMIA.

The positive diagnosis of these conditions can only be secured by a microscopic examination of the blood.

In pernicious anæmia, according to Dr. EICHHORST, the characteristic appearances are : A portion of the red corpuscles are seen to retain their normal size, but are marked by an extreme paleness, with a tendency to crenation and the formation of rouleaux, while others among them attract attention by their small size, which is reduced often to one-fourth the diameter of the well-formed corpuscles. These small ones are more deeply colored, and if allowed to roll over under the thin cover-glass, their appearance in profile shows them to have lost to a greater or less extent their bi-concave outline.

Dr. F. P. Henry* has found in the blood of pernicious anæmia

---

* *Phila. Medical Times*, April 3, 1886.

a very interesting illustration of "tissue reversion." The red blood corpuscles are occasionally seen with nuclei, and resemble the corpuscles of the blood of cold-blooded animals in all their principal characteristics; namely, in "their number, their size, their shape, and the amount of hæmoglobin they carry."

For the examination of the blood in such investigations, Dr. Gowers, of London, recommends the use of the *hæmacytometer*, by which he measures for the purpose of ascertaining the number of red and white cells in a given volume of blood. The essential part of the apparatus consists of a glass slip, on which is a cell one-fifth of a *millimètre* (.008 inch) deep. The bottom of this cell is divided into one-tenth *millimètre* squares. Upon the top of the cell rests the glass cover, which is kept in its place by the pressure of two springs. In estimating the number of corpuscles, the patient's finger is pricked; then by means of a capillary pipette, five cubic *millimètres* of blood are taken up and well mixed up with 995 cubic *millimètres* of saline solution; a drop of the dilution is then placed in the glass cell, the cover is adjusted, and the slide is placed in the field of a microscope. In a few minutes all the corpuscles have sunk to the bottom of the cell, and are seen lying on the squares; the number of corpuscles in ten squares is then counted, and this, multiplied by 10,000, gives the number in a cubic *millimètre* of blood. The degree of dilution and size of the squares are so proportioned that, with normal blood, two squares contain about 100 corpuscles, and the number in two squares thus expresses the percentage proportion of corpuscles to that of health. The proportion of white corpuscles to red, or their absolute number, may be easily determined during the same observation. The Globule Counter of MALASSEZ is believed to furnish more exact results than that of GOWERS.*

A simpler method was used by Dr. J. G. RICHARDSON, of Philadelphia. He spread a drop of fresh blood thinly on a glass slide, letting it dry, and then counted the number of white corpuscles. The specimens when thus prepared can be kept dry for any length of time, if preserved from dust and moisture, so that by comparing specimens of

---

* For a description see *Archives de Physiologie* for June, 1880, or Da Costa's Medical Diagnosis, 6th Ed., Phila., 1884, page 752.

different persons' blood, prepared similarly, the variations in the number of white corpuscles can be readily observed. By this means he claimed to be able to detect leukæmia in its early stages.

Profound anæmia is met with in the following conditions: (1) After great loss of blood or exhausting discharges; (2) where there is inanition (insufficient nourishment); (3) in chlorosis; (4) in malignant disease; (5) in Bright's and Addison's disease; (6) leucocythemia; (7) chronic metallic poisoning, and (8) in malarial toxæmia.

The symptoms of the idiopathic or "progressive pernicious" form of anæmia are described by Dr. BYRON BRAMWELL, as follows: A profound anæmia, which is associated with marked changes in the microscopical characters of the blood, and (in most cases) with the presence of retinal hemorrhages. The patient is generally well covered with fat, the skin is smooth and soft, the face looks slightly swollen, and is of a pale yellow or yellowish-green color. All the symptoms of profound anæmia are present, viz., extreme pallor of the mucous membrane, great debility, tendency to fainting, dyspnœa and palpitation on exertion, buzzing in the ears, headache, subcutaneous œdema, etc.; loud blowing murmurs are heard over the heart and great vessels; there is a venous hum in the neck; the pulse is very soft and compressible. Attacks of vomiting and diarrhœa are frequent; irregular elevations in temperature, transient paralyses, hemorrhages from the mucous membranes, occasionally occur. The causes of the disease are at present unknown. The disease is said to occur more frequently in women than in men. In the majority of cases the termination is in death, the end being ushered in by profuse diarrhœa, coma, or delirium. The diagnosis is facilitated by examination of the blood.*

Great advances have been made within a few years in the department of microscopic botany, and sufficient evidence has been accumulated from experiment and observation, to warrant the statement that many diseases both in man and the lower animals are due to infection by micro-organisms, termed collectively microbes (*microbies*) by Pasteur. The methods pursued in studying this interesting and very important question of the relation of these organisms to disease, are of

---

*See article on Diagnosis of Progressive Pernicious Anæmia, by Dr. F. P. HENRY, *Phila. Medical Times*, vol. xvi., page 499.

three kinds, viz.: (1) By microscopic manipulation (behavior to certain reagents and coloring fluids, etc.); (2) By isolation and culture of
the morbid agent (on gelatine, agar-agar, peptone, and in various
other media); and (3) By inoculation of the pure cultures, or the secretions obtained originally from typical cases of disease.

It has been found that suppuration is due to the presence in the tissues of three varieties of staphylococcus pyogenes (aureus, citrus, or
albus) and the streptococcus pyogenes.   In septicæmia in man, Klein
has found in the blood-vessels of the swollen lymphatic glands, large
numbers of minute bacilli slightly thicker, but otherwise resembling
the organisms claimed by Koch as the cause of septicæmia in mice.
These organisms do not originate within the body, but are introduced
from without (infection), hence the explanation of the care used to
keep wounds clean in the antiseptic dressings; precautions which are
fully justified by the wonderful achievements of modern surgery.

Among internal diseases, relapsing fever has been long known to be
accompanied by a spirillum in the blood, as pointed out by Ober-
meyer in 1873.   It is the only form of spiro-bacterium which is known
to be pathogenic to man; all the others belong either to the class of
micrococci (round bodies) or bacilli (rod-shaped bodies).   Micrococci
have been found in the pus of open wounds, and in closed abscesses
as well, occurring singly and in chains or zoogloeic masses.   In all
cases of diarrhœa, the discharges from the bowels swarm with micrococci, and in typhoid fever they are also found colonized in the ulcers
and also in the neighboring mesenteric glands.   Wassilieff has shown
that these micrococci only occur after the death of the tissue or tissues, so that in these they may multiply so as to form extensive col-
onies, and that therefore the presence of these micrococci is only a
secondary phenomenon.*

Micrococci also occur in tuberculosis, and in severe catarrhal pneumonia, presenting a close analogy between these diseases and pleuro-
pneumonia of cattle or the pneumonia of swine fever.   Micrococci are
also found in normal human saliva, and STERNBERG considers that they
are the active agents in causing septicæmia when saliva-injections are
made into rabbits, but this is not considered as proved.†

* *Centralblatt für die Med. Wissen.*   1881.
† Klein.   Micro-Organisms and Disease.   London, 1884.

On the other hand, CHAUVEAU, COHN, WEIGENT, POHL-PINCUS and others have apparently shown that the morbific agent in small-pox and in vaccinia is a micrococcus; FEHLEISEN claims that erysipelas is due to the *micrococcus erysipelatosus*, while BUHL, HUTER and OERTEL claim that the micrococcus diphtheriticus is the cause of the manifestations of diphtheria. This, however, has been denied by WOOD and FORMAD, who see nothing specific in the micrococcus found in the grey pseudo-membrane of diphtheria. NEISSER has detected a micrococcus gonorrhœæ which has been called for brevity rather than elegance, a "gonococcus." Similar organisms have been found in ulcerative endocarditis, in scarlatina, and in puerperal fever. FRANKENHAUSER discovered a micrococcus in the blood of pregnant women suffering with pernicious anemia; AUFRECHT describes them as occurring in syphilitic mucous patches; SCHULER and ROSENBACH, in the lesions of acute infectious osteo-myelitis; but their specific differences, if they exist, as well as distinguishing characteristics, remain to be definitely defined.

L. DOMINGOS FREIRE ascribes yellow fever to the micrococcus xanthogenicus, and he has made an extended series of inoculation experiments for the prevention of the disease, with reported success. Dr. CARMONA, of Mexico, also claims similar success, but it is too early to decide either upon the active agent or the accuracy of the reports which have been made.

Bacilli present more distinctive forms and are more susceptible to investigation. The pathogenic forms in man are numerous. The bacillus septicæmia of KLEIN has been already referred to. KLEBS and EBERTH claim that in the bacillus typhosus exists the infective principle of typhoid fever. The bacillus malaria of KLEBS and TOMMASI CRUDELLI has been further investigated by LAVERAN, and the opinion of more recent investigators is that the "malarial body" is less a bacillus than an infusorian.*

ARMAUER HANSEN (Virchow's Archiv. B. lxxix.) first ascertained the existence of a peculiar bacillus in the large leprosy cells of VIRCHOW, which has been named the *bacillus lepræ*. The *Bacillus anthracis*

* See page 46 for fuller consideration of the relation of this to malarial toxæmia.

occurs in charbon or wool-sorter's disease (DAVAINE, BOLLINGER, KOCH and others).

KOCH finds in the bacillus tuberculosis the efficient cause of all tubercular lesions.* Similar elements have been found in tuberculosis of the skin (lupus). *Comma shaped bacilli* are declared by KOCH to be the cause of chôlera, but his conclusions are still under discussion, and some of his statements are contradicted by other investigators who are now engaged in studying this important subject.

The purulent joint inflammation with metastatic abscesses and re-peated chills of pyæmia will prevent its being taken for rheumatism. Gonorrhœal rheumatism is regarded as a mild form of pyæmia. The distinction usually observed between pyæmia and septicæmia is based upon the following·differences: †

PYÆMIA.	SEPTICÆMIA.
Commonly commences with a chill.	Commonly commences without a chill.
Fever variable, but rarely entirely intermits.	Fever steadily increases, but is lower in the morning.
Sudden and great changes in temperature, followed by profuse perspiration.	The temperature is high at the beginning of the disease, increases usually near the fatal termination, when it falls below the normal. The skin is moist, but without profuse sweating.
Pulse variable; toward the fatal end, rapid, feeble, and irregular.	Rapid and gradually increases in frequency toward the latter end.
Facies, at the beginning flushed or pallid, toward the end, careworn.	Expressive of a dull, listless condition throughout the whole course of the disease.
Tongue smooth, dry, and excessively red; later brown, and even the teeth covered with sordes.	Tongue, lips, and throat dry at the commencement; toward the end, moist. Thirst is marked.
Diarrhœa, with stools of a pappy consistence.	Rice-water evacuations, very offensive; obstinate vomiting.

* For method of staining, etc., see page 65.

† B. A. Watson, article on Pyæmia and Septicæmia, in Pepper's System of Medicine.

PYÆMIA. ·	SEPTICÆMIA.
Epistaxis.	Epistaxis rarely occurs.
Mild delirium towards the fatal end.	A lethargic condition from the beginning, increasing toward the fatal end.
Aphthæ in the mouth and throat, sudamina, vesicles, pustules, and purpuric patches.	Icteric hue of conjunctivæ; singultus often present.
Pus and blood contain globular bacteria.	Secretions and blood contain red bacteria.
Secondary wound complication rarely develops before second week after receipt of injury.	Primary wound complication generally developed within forty-eight hours.
Increased coagulability of the blood.	Diminished coagulability of the blood.
Metastatic abscesses common.	Complete absence of purulent or ichorous deposits in unmixed septicæmia.

PART II.

LOCAL DISEASES.

CHAPTER I.

DISEASES OF THE NERVOUS SYSTEM.

CONTENTS.—*Cerebral Disorders—Congestion—Anæmia—Apoplexy—Thrombosis—Embolism—Meningitis—Tubercular Meningitis—Rheumatic Meningitis—Acute Cerebritis—The Ophthalmoscope in Nervous Disorders—Headache—Chronic Cerebral Disorders—Hypertrophy—Hydrocephalus—Brain Tumor—White Softening—Abscess—Chronic Meningitis—Thrombosis—Sclerosis—Localization of Brain Disease—Lesions of Cerebral Cortex—Brain Lesions other than Cortical—Tabular View of Paralysis with Seat of Lesion—Spinal Disease—Organic and Functional Paraplegia—Diagram of Spinal, Inflammatory and Degenerative Diseases—Tabular View of Spinal Paralysis, Congestion, Meningeal Apoplexy, Spinal Apoplexy, Acute Primary Myelitis—Comparison of Acute Spinal Disorders—Tumors—Tremors—Chronic Degenerative Diseases of the Cord—Prof. Charcot's Diagnostic Chart of Cerebro-Spinal Affections—Patellar-Tendon Reflex of Westphal—Analysis of Symptoms of Focal Lesions of the Cord (Gowers, Charcot, and Erb)—Comparative Semeiology of Cerebro-Spinal Sclerosis, Paralysis Agitans, and Locomotor Ataxia—Paraplegia from Reflex Irritation and Myelitis Compared—Gowers' Classification of Spinal Lesions—Pseudo-Hypertrophic Paralysis—Lead Palsy—Hysterical Paralysis—General Paralysis of the Insane—Spinal Irritation, and Spinal Weakness—Hysteria and Hystero-Epilepsy—Neuralgia—Insanity, Its Different Forms, their Pathology and Etiology.*

Recent advances in the physiology of the nervous system have thrown much light upon mental and nervous manifestations; and many

6   · (81)

conditions which had been hitherto regarded as primary have been shown to be in reality symptomatic and secondary to definite morbid changes occurring either in the central nervous system, the trunks of the nerves, or in the peripheral terminations. It is evident that diseases having their origin or seat of lesion in the nervous system will vary in their symptoms with the locality of the morbid process and the function of the part affected. . Disorders of intellection and insanity result from involvement of the cerebral hemispheres, especially of their anterior portion, with impairment of special senses, and paralysis of parts supplied by cranial nerves, occurring with or without loss of power in the extremities. Diseases of the spinal cord give rise to paralysis of muscles having direct connection with the seat of lesion, and also to disorders of sensation and nutrition. Hemiplegia may be of cerebral origin; paraplegia is generally spinal. Pressure or irritation of nerve trunks may cause local palsy, spasmodic affection, or neuralgia, while myopathic paralysis (such as encountered in lead palsy, pseudo-hypertrophic paralysis, and progressive muscular atrophy) may be due to a peripheral nervous affection. Hysteria, vertigo, neurasthenia and some mental disorders, being of uncertain seat and unknown relations, may be provisionally considered as functional disorders of the nervous system.

The principal symptoms referable to the brain may be considered as being caused by (*a*) congestion, anæmia, apoplexy, thrombosis, embolism, brain tumor, cerebritis and abscess; (*b*) by influence upon the brain disease of neighboring structures, such as meningeal inflammation, hemorrhage, effusion or neoplasm, necrosis, disease of the middle ear; and by (*c*) poisoned conditions of the blood, as in uræmia, alcoholism, and the delirium of fevers. An irregular and abnormal distribution of the blood supply may give rise to night terrors, epilepsy, syncope, temporary (functional?) paralysis, cerebral exhaustion, migraine, aphasia, aphemia and agraphia; and irregular motor discharges from the cerebral centres are directly associated with chorea, tremor, and epileptiform convulsions; the higher mental powers being apparently merely held in abeyance in catalepsy, trance and hysterical coma.

Passing to the diagnosis of the principal cerebral disorders, the following points are of importance in distinguishing cerebral congestion and cerebral anæmia:

CEREBRAL CONGESTION.		CEREBRAL ANÆMIA.
Severe, throbbing and diffused.	Headache.	Less sharp, generally vertical.
May be absent.	Vertigo.	Usually a marked symptom.
Full, throbbing, tortuous and distinct.	Temporal Vessels.	Not prominent.
Pulse full, tense; often signs of plethora.	General Circulation.	Pulse irritable, often anæmic murmur of pulmonary artery.
May be rumbling or singing.	Tinnitus aurium.	Noises may be short and high-pitched.
Hallucinations; may have active delirium.	Mental Phenomena.	Below normal; incapacity for continual mental application.
Surface temperature of scalp may be increased.	Temperature.	Surface temperature, if at all affected, is diminished.
Contracted.	Pupils.	Dilated.
Not increased; may contain urates and phosphates (HAMMOND).	Urine.	Limpid, and may be passed in excess; decrease of salts.

Cerebral exhaustion is sometimes so marked as to produce coma, and thus form a variety of apoplexy, and its diagnosis is made by excluding hyperæmia, hemorrhage, embolism and thrombosis of basilar artery. Where the latter condition terminates in recovery, it is almost identical in its manifestations; indeed, " it would be difficult to disprove the assertion that cases of cerebral exhaustion belong in this category" (FLINT).

A sudden attack of coma in a case of albuminuria may be set down as due to *uræmia*, if embolism and apoplexy are excluded (by noting the absence of hemiplegia). Should the coma be associated with epileptiform convulsions, this diagnosis is likely to be correct, even if no albumen can be detected in the urine; since the form of renal disease most likely to give rise to uræmic poisoning is the cirrhotic form (contracted kidney), in which the albumen may be absent from the urine for considerable periods of time.

The early diagnosis of diseases attended by coma is of great importance, with the view of promptly instituting proper treatment.

CEREBRAL APOPLEXY.

Apoplexy must be distinguished from drunkenness, narcotic poisoning, uræmic poisoning, epilepsy, concussion of the brain, cerebral
thrombosis, embolism, and insolation or sunstroke.

Drunkenness. The odor of liquor may excite suspicion. If the
patient vomit, the ejecta may be tested for alcohol. Or the urine may
be examined by Anstie's test, as follows:—

R.	Bichromate of potash,	1 part	
	Strong sulphuric acid,	300 parts.	Mix.

To fifteen minims of this add a few drops of the urine, and if the
patient has taken a toxic dose of alcohol, the mixture will turn an
emerald green. In drunkenness, the pulse is generally rapid, the
pupils not dilated, the eye injected. The patient can be roused, and
hiccoughs.

A modification of this test, introduced by WOODBURY, is to put in a
small test-tube a cubic centimetre of colorless sulphuric acid, and add
an equal quantity of urine, so as to form a layer of urine over the acid.
A crystal of bichromate of potassium is now added, and the liquid
slowly mixed by rotation. If a proportion of alcohol amounting to
three parts in a thousand be present a bright green color will be formed,
otherwise the solution remains a light red color.

Dr. MacEwan, of Glasgow, gives the following method of distinguishing alcoholic coma from that of apoplexy, fracture of the skull,
and other causes. In alcoholic coma, as long as the patient is undis-
turbed the pupil is contracted; but if any stimulus not sufficient to
arouse the patient be applied to him, such as a shake or a pull of the
beard, *the pupil dilates,* only, however, to become contracted again as
soon as the person is left at rest.*

Narcotic poisoning. In this condition the outset is gradual; there are
often convulsions, but the patient may be roused. In opium poisoning the pupil is extremely contratcted; so it is in hemorrhage in the
pons. The vomiting, the acrid odor of opium, and the gradual intensification of the coma, are diagnostic. There is no hemiplegia.

Uræmic poisoning. Here the coma nearly always comes on grad-

British Medical Journal, November 16, 1878.

ually and is preceded by general convulsions. It is not deep, and at first the patient may be aroused. The stertor of the breathing is more superficial, while there is also frothing at the mouth.

Nearly always, distinctive modifications of the heart-sounds will be heard, as reduplication of one or both, intensity of second sound, etc.; while there are elevation of the arterial tension and increased cardiac impulse. Of these cardiac physical signs none seem so constant or remarkable as muffling of the first sound. (Mr. W. WHITTLE.)

There are, moreover, in many cases marked prodromata. The skin has been waxy and œdematous, the eyelids puffed, and the legs and feet swollen. The urine may or may not be albuminous (but albumen may also be present in apoplexy).

Epileptic coma presents a history of convulsions; lasts but for an hour or two; there is frothing at the mouth; and the temperature is elevated.

In *hysteria* and *catalepsy* there is no alteration of temperature, and no frothing at the mouth.

In *concussion or compression* from injuries to the head, the skin is pale, the pupil dilated, and vomiting occurs. The symptoms are usually of short duration, and there is usually a history of injury. Meningeal hemorrhage from injury presents no points of difference from true apoplexy, except that hemiplegia is generally wanting (FLINT).

Syncope is readily distinguished by the feeble pulse, the pale face, the quiet respiration, and the brief duration of the unconsciousness; while in *asphyxia* the livid face, distressed breathing, and blue lip; which precede the coma, indicate its distinction.

REFLEX HEMIPLEGIA.

The details of a case are given in abstract in the *Practitioner* from M. Barbez, who reports it in the *Revue de Médecine* for June, 1886, in which five weeks after the evacuation of a highly-offensive empyema the patient became suddenly paralyzed in the right arm, with aphasia, and continued for more than two months in this condition, with mild delirium. There were some spasms of the affected arm.

Especial attention is attracted to the case by the explanation which is given of the connection between the two conditions. M. Luys considered it a case of limited left lateral meningitis. But M. Barbez classes it among so-called *reflex* hemiplegias, cases of which have been reported by several writers. The explanations seem very far-fetched to us. As the patient had a good ground-work for pyæmia, we would rather suppose that his brain-trouble was of septic origin, probably embolic.

In regard to *thrombosis* and *embolism* of the larger cerebral vessels, the diagnosis is often extremely difficult. The following table of the comparative symptoms is drawn up from the works of BUDUY, GELPKE, FLINT, and HAMILTON:

CEREBRAL HEMORRHAGE.	CEREBRAL THROMBOSIS.	CEREBRAL EMBOLISM.
Occurs most frequently in advanced age, with atheromaious arteries.	In advanced age. May occur in children during scarlet fever and renal disease.	Almost always in early or middle life (FLINT).
Onset generally sudden.	Development of symptoms gradual.	Onset rapid, without premonition.
Hypertrophy of left ventricle. Alcoholism, or other debilitating habits.	No rheumatic history. Endarteritis deformans of peripheral arteries sometimes present.	Previous articular rheumatism or other diseases leading to formation of clots. Often cardiac valvular insufficiency. Coincident embolisms are sometimes present elsewhere in the body.
Pain in the head.	No headache.	No headache.
Ataxic aphasia, secondary to loss of consciousness. Intelligence much affected.	Aphasia incomplete and primary, occasionally absent. Intelligence less involved.	Amnesic aphasia. Retention of mental power.
Often coma.	Rarely loss of consciousness.	No coma.
Paralysis very marked; occurs on either side.	Paralysis less marked.	Muscular paralysis extensive; nearly always on the *right* side (FLINT).
Apoplectic phenomena from the outset. Symptoms of cerebral pressure.	No apoplectic phenomena at onset.	Early apoplectic phenomena, but without loss of consciousness.
Disappearance of the residual disorder after a moderate time. May terminate in chronic abscess.	Recovery slow; more or less hemiplegia may remain.	Very rapid, or else quite imperceptible disappearance of the residual disorder. May be followed by softening.
After a few days pain in the head and increased temperature of the body on the *unaffected* side (FLINT).	May have œdema more marked on affected side.	One-sided œdema, often in the arm alone.

The high temperature (108° to 113° F.) of cases of sunstroke serves to distinguish such from the coma of apoplexy; although in some cases of insolation the coma is probably due to cerebral exhaustion without high bodily temperature, the distinguishing features of which have been previously considered. The subjects attacked are generally laboring men, who have been exposed, while at their work, to a continuous high temperature.

ACUTE CEREBRAL INFLAMMATION.

Considering the acute inflammatory state of the brain and its coverings, we tabulate their comparative semeiology as follows:

SIMPLE MENINGITIS. (*Lepto-meningitis.*)	TUBERCULAR MENINGITIS.	RHEUMATIC MENINGITIS.	ACUTE CEREBRITIS AND CEREBRAL ABSCESS.
Due to disease of the cranial bones, traumatism, exposure to sun. (Very frequently the meningitis of young adults has a syphilitic source.) May be epidemic.	Scrofulous inheritance.	Rheumatic history or diathesis.	May be due to general causrs, such as pyæmia, etc., or to local causes, as traumatism, bone disease, local irritation, extension from meninges, etc.
A disease of both infants and adults, though usually in the latter.	Often children under five years of age.	Adults.	Often in elderly subjects.
Previously healthy; no prodromata.	History of persistent headache and obstinate constipation; wasting.	Often during an attack of joint inflammation.	Rarely occurs in previously healthy persons.
No chest symptoms.	Previous pulmonary trouble.	None.	None.
Onset sudden.	Takes four or five days to develop; approach insidious.	Rapidly developed.	Slow, and may simulate typhoid.
Headache intense on both sides of head.	Persistent and marked headache, which exacerbates.	Intense pain.	Dull, persistent and localized; less than in meningitis.
Pupils contracted.	Pupils irregularly dilated.		
Intelligence clear at first, but may become furiously delirious.	Delirium of low grade at night (stupor in second stage); strabismus, and oscillation of eyeballs.	Leads to active delirium.	Mental confusion and impairment of intelligence.

SIMPLE MENINGITIS. (*Lepto-meningitis.*)	TUBERCULAR MENINGITIS.	RHEUMATIC MENINGITIS.	ACUTE CEREBRITIS AND CEREBRAL ABSCESS.
Vomiting early, frequently.	Vomiting occasionally.	Not marked.	Vomiting not infrequent.
Pulse full and rapid.	Irregular and slow pulse.	Pulse full and rapid.	
High fever.	Fever not intense.	Temperature may be very high.	Less fever.
Convulsions early, contracted pupils, with contractions of flexor muscles of arm or leg.	Convulsions late, with dilated pupils and hemiplegia.	No convulsions.	No convulsions; but sudden hemiplegia may occur.
In fatal cases death generally occurs in a week; recovery is slow.	Lasts from one to three weeks.	Lasts a few days; death often occurs from continued high temperature.	Course is often chronic.
Prognosis favorable under prompt treatment.	Prognosis unfavorable.	Prognosis fair.	Prognosis not encouraging.

Dr. GEE notes that meningitis of the base of the brain is generally tubercular; and when tubercular meningitis attacks the convexity, there is a constant convulsive condition, moderate force and very variable pulse. (See page 52 for a more detailed account of tubercular meningitis.)

These cerebral diseases may be distinguished from typhoid fever by the history and course of the affection. Typhoid occurs in the spring and fall, and is often endemic; it rarely appears in children, and generally attacks young adults. It is a continued fever, coming on in a hitherto healthy person with malaise, epistaxis and diarrhœa. Chills and vomiting are rare. Convulsions and paralysis, if they occur at all, are late manifestations, and due to complications. Delirium is of low type, headache dull, moderate deafness, pulse rapid but regular or dicrotic. Abdominal symptoms generally prominent, tympanites, diarrhœa, tenderness and gurgling on pressure in the right iliac fossa, and a discrete rose-colored eruption upon the chest and belly. Convalescence at the beginning of the third week; disease generally continues about four weeks.

THE OPHTHALMOSCOPE IN NERVOUS DISORDERS.

In the diagnosis of intracranial disorders the ophthalmoscope is often of great service, though, perhaps, scarcely to the extent advocated by BOUCHUT. The *discrete tubercles of the choroid* accompanying meningeal deposit, the *choked disc* in cerebral tumors and inflammations, and the *retinitis* and *retinal hemorrhages* of Bright's disease are of great importance. BOUCHUT declares* that the ophthalmoscope is as indispensable to the physician as to the oculist, and he was among the first to point out the great importance of this aid to practical medicine. We quote his opinions and conclusions:

"All diseases of the brain and spinal cord, and all the nervous affections termed neuroses, because they are regarded rather as functional than organic, ought to be investigated by its aid. When by its assistance the physician discovers a lesion of the optic nerve, of the retina, or of the choroid, in a case presenting convulsive, choreic, paralytic, or spasmodic nervous phenomena, he may be certain that a cerebrospinal lesion is the starting point of these symptoms. Every symptom regarded as nervous, which is accompanied by a lesion of the fundus of the eye, is caused by an organic alteration of the brain, the cord, or the membranes. Thus is it with chorea, considered by many physicians as a simple neurosis; and yet this should, in consequence of the congestive optic neuritis found in its subjects, be regarded as a congestive affection of the anterior spinal columns. So also epilepsy, in a certain number of cases, is the result of cerebro-spinal lesions which at the same time induce changes in the optic nerve or retina. Hysterical paraplegia and paralysis produce no neuro-retinian changes, while symptomatic paraplegia and spinal ataxia produce either simple hyperæmia of the optic nerve or hyperæmia and atrophy. So leucæmia, tubercular, glycosuric, or albuminuric diathesis are often revealed by optic neuritis, the ophthalmoscopic diagnosis in some of these cases being most striking. It is especially in patients attacked by general acute tuberculosis, accompanied by typhoid symptoms, and which are mistaken for typhoid fever, that cerebroscopy becomes truly remarkable. In an infant in whom the disease had all the appearance of

* "Revue Cérébroscopique," in *Gazette des Hôpitaux*, for January, 1874.

typhus, the ophthalmoscope, by revealing tubercles of the choroid with neuro-retinitis, determined that there were tubercles in the brain, and consequently productions of the same character all over the body—which the autopsy demonstrated to be the fact.

"Can any diagnosis be more exact than this? You see, in the living man, tubercles of an organ which permit you to conclude that they will also be found elsewhere. You see a nerve either healthy or diseased, and this indicates whether its roots are sound or diseased; and you have almost laid bare arteries and nerves which are so afferent to the brain that changes in them, studied with care, represent similar changes in a portion of the nervous centres. It seems almost marvelous; and I do not think that since auscultation there has been anything discovered so useful to semeiology. Henceforth, the physician may divine and often affirm lesions of the brain, cord, or meninges, the diagnosis of which before was impossible or only probable. Thus: 1. From hyperæmia and hyperæmic tumefaction of the optic nerve there results the diagnosis of mechanical or inflammatory hyperæmia of the brain in meningitis, in cerebral hemorrhage, effusions into the brain, and in some cases the diagnosis of ataxic or other spinal diseases. 2. By papillary œdema joined to hyperæmia I recognize œdema of the meninges; or an obstructed cerebral circulation through meningitis, cerebral tumors, ventricular hydrocephalus, cerebral hemorrhage, meningeal effusions, thrombosis of the sinus, etc. 3. By neuro-retinian and choroidean anæmia, I recognize cerebral hemorrhage of *ramollissement*, and if the anæmia be absolute it is fatal. Empty arteries and veins of the eye, and an exsanguineous condition of the choroidean network, indicate arrest of cerebral and cardiac circulation. 4. By exudative and fatty optic neuro-retinitis, I recognize chronic meningo-cephalitis; the encephalitis of cerebral tumors, and the changes in the nervous substance which accompany these tumors. 5. By retinian varices and thromboses, I distinguish meningeal thromboses, or those of the sinuses. 6. By the aneurisms of the retinian arteries we may recognize the miliary aneurisms of the brain. 7. By simple retinian hemorrhages we recognize a compression of the brain by hemorrhagic or other effusions; but if these retinian hemorrhages are accompanied by retinian steatosis, there is also cerebral steatosis,

and this is the case in chronic albuminuria, leucocythæmia, and glycosuria. 8. By atrophy of the optic nerve, tumors of the brain and cerebral or spinal sclerosis are discovered. 9. Finally, we never meet
with tubercular granulations in the choroid without the existence of
similar ones in the lungs or other organs."*

The ophthalmoscope is now frequently employed for diagnostic
purposes in ordinary medical practice where there is imperfection of
vision, to determine whether it is due to other than nervous lesions, to
discriminate between affections of different portions of the eye, and,
sometimes, to measure the amount of refraction in cases of hyperme-
tropia and myopia. Even where there is no impairment of sight,
there still may occur very decided and characteristic retinal changes
and alterations in the optic disc, which are readily detected by ophthalmoscopic examination, as already indicated; so that in obscure
cases the routine examination of the eyes has become nearly as im-
perative as the chemical and microscopical examination of the urine.

HEADACHE.

Some of the most trying cases to treat are those of headache, because this symptom may appear in many and even diverse morbid
states, and often indicates serious cerebral disorder. Mr. WM. HENRY
DAY, of London,† has made a study of these conditions, and thus .
summarizes his conclusions :—

Headache usually denotes some functional disturbance of the brain
or its membranes, induced (1) by excess of local blood pressure, (2)
by absorption into the blood of poisonous matters, (3) or by such a
diminution of healthy blood as provokes irritation and suffering. It
may be a symptom of organic disease, either of the brain or its membranes, or of the kidneys or stomach, and uterus.

Cerebral anæmia.—A striking symptom is pain at the *top* of the
head, which often feels hot and burning, sometimes gnawing and
scraping. Irritability of temper. Face livid and cold. Patient easily
exhausted. Eyes dull.

Hyperæmia, Active or Passive.—Active.—Arterial fullness. Head hot,

---

* *Medical Times and Gazette,* January 23, 1875.
† *British Medical Journal,* Nov. 16th, 1878.

pain frontal, throbbing and bursting, pulse tense, full. Conjunctivæ reddened. Eyes bright. Photophobia. Mentality dull. Apoplexy may ensue.

Passive.—Venous fullness from obstruction caused by heart disease, bronchocele, etc., pleuritic effusion, defective ventricular action.

Sympathetic headache.—Faulty digestion or ovarian excitement. Stomach sometimes weak and over-sensitive. Catamenial headache. Dyspeptic and bilious headache. *Irritation of the sympathetic reduces the amount of blood in the brain.*

Nervous headache.—Disturbance of brain from overwork, worry and anxiety. Aggravated by some of the circumstances favoring sympathetic headache. In women there is a passage of a large quantity of limpid urine; feet and hands cold. Confusion of ideas. Nausea and sickness, not attributable to errors in diet, may precede the attack. In nervous people constipation may cause headache.

Poor seamstress headache.—Spanæmia. Headache of excessive menstruation, or menorrhagia. Hereditary influence strong.

Neuralgic headache.—From decayed teeth, peripheral irritation, malarial poison. Pain and tenderness along the fifth nerve. Pain intense; not relieved by vomiting.

Toxæmic headache.—Poisoned blood acting on nerve centres, from particular articles of food or drink, and drugs. Or certain specific diseases—gout, rheumatism, and .syphilis. Headache of vitiated atmosphere. Tea or coffee headache.

Organic headache.—Morbid growths; meningitis. When slowly progressing pain is limited to smaller area, and is intense. Periosteal inflammation is accompanied by tenderness upon pressure.

Headache in children, due to accidental injuries, to derangement of alimentary canal, anæmia, exhausting influences, such as bad food and impure air, immoderate intellectual efforts, and sometimes to organic diseases (often tuberculosis).

In strumous and weakly children headache must be carefully watched. A headache of long standing in a child is significant, and requires more serious attention than in the adult.

CHRONIC CEREBRAL DISORDERS.

In children the diagnosis may be required to be made between hypertrophy of the brain and hydrocephalus, which have enlargement of the head as a common sign.

HYPERTROPHY.	HYDROCEPHALUS.
Increase in size most marked above the superciliary ridges.	Increase in size most marked at the temples.
Head square in shape.	Head more rounded.
No yielding of fontanelle on pressure.	Fontanelle elastic.
Eyes at normal distance.	Distance between the eyes increased.
Excessive amount of brain, especially white matter.	Excessive amount of fluid in ventricles, or sub-arachnoid space.
Patient dull, liable to epileptic fits, and suffers from headache.	Mentality feeble; generally can be traced to congenital source; death may occur from convulsions. No marked headache.

The diagnosis of hydrocephalus may be confirmed by tapping the fontanelle with the aspirator, or a hypodermic syringe. In adults brain tumor and sclerosis are among the prominent disorders of slow progress, the symptoms varying in a very marked manner with the location of the lesion. Chronic inflammation of the brain may terminate in insanity or in abscess. In its course it has been mistaken for dyspepsia, but a proper inquiry into the mental condition of the patient will reveal the cerebral mischief, which continues to progress even after any coëxisting indigestion has been corrected. There is, moreover, sluggish intelligence, and partial paralysis or rigidity of certain muscles of the extremities. Attacks of delirium or mania finally confirm the diagnosis, and the patient usually dies in a state of coma.

Intra-cranial disease of a chronic character is often so obscure as to leave even the most experienced in doubt, and the post-mortem examination sometimes produces revelations that disconcert the medical attendant. Due regard to some of the characteristic phenomena in the accompanying table will often serve to clear up the doubts surrounding a difficult case.

Brain Tumor.	Softening (White).	Abscess.	Chronic Meningitis.	Thrombosis of Sinuses of Brain.
Of slow development.	Approach and progress slow. Follows embolism or apoplexy. Non-inflammatory.	Follows injury to the skull, or chronic disease of the head.	Caused by syphilis, rheumatism, disease of bones, blows upon the head, etc.	Sudden development of symptoms.
Intellect not disordered at first.	Early affection of intelligence. Marked impairment of memory.	Varies with seat.	Intelligence not affected, except during attacks of delirium.	May be unconsciousness or not. Intelligence subsequently good.
Headache violent, paroxysmal and often localized.	Dull and constant.	Sudden in its development and general.	Subject to exacerbations, but generally chronic.	No headache; œdema of forehead and eyelids.
Paralysis slow in appearing, and often limited to the muscles of eye or of the face; more rarely hemiplegia.	Motor and sensor phenomena more frequent and prominent. Partial palsies and disturbances of sensibility subsequently.	Course is much more rapid; convulsions, drowsiness, paralysis and coma quickly developed.	In consequence of meningeal exudation may present the clinical signs of a brain tumor.	May be coma; varies greatly, according to part of brain whose vascular supply is disturbed.
Convulsions a common symptom, epileptiform in character. Not followed by palsy or hebetude.	Begins often with apoplectiform attacks, which seldom occur afterward.	Convulsions early; paralysis belongs to developed stage.		Very rare.
Vertigo and tinnitus aurium.	Vertigo.		More vertigo.	Varies.
Vomiting.	Not unfrequently.	Rare.	Frequent vomiting.	No vomiting.

SCLEROSIS.

Sclerosis is a disease of the nerve centres, in which there is increase of connective tissue elements, without primary involvement of the nerve cells. It may exist as diffused cerebral sclerosis, spinal sclerosis (several forms), cerebro-spinal sclerosis (sclérose en plàque), and glosso-labio-laryngeal paralysis. Cerebral sclerosis occurring in children can be distinguished from deficient development by the following characteristics:

DEFECTIVE DEVELOPMENT OF INTELLIGENCE.	DIFFUSED CEREBRAL SCLEROSIS.
Intelligence stationary, instead of progressing with age.	Intelligence retrogressive and more affected. Often terminates in idiocy.
Not connected with disease.	May follow injury to head, zymotic fevers, severe·application of body or mind.
Speech restricted to few words, imperfectly pronounced.	Never learns to talk, or speech becomes imperfect or lost after it has been acquired.

DEFECTIVE DEVELOP-MENT OF INTEL-LIGENCE.

No paralyses.

Muscular system in good condition.

No convulsions.

Improved by training and education.

DIFFUSED CEREBRAL SCLEROSIS.

Usually more or less hemiplegia.

Arrest of growth of certain parts of body, with contraction and distortion of affected limbs.

Frequent convulsions.

Progress very chronic, and may live to advanced age.

LOCALIZATION OF BRAIN DISEASE.

The localization of diseases of the brain is a subject of great interest. In order that a correct diagnosis should be made, the important anatomical and physiological data must ever be borne in mind. We proceed first to the consideration of

LESIONS OF THE CEREBRAL CORTEX.

[The accompanying excellent diagram, or physiological map of the

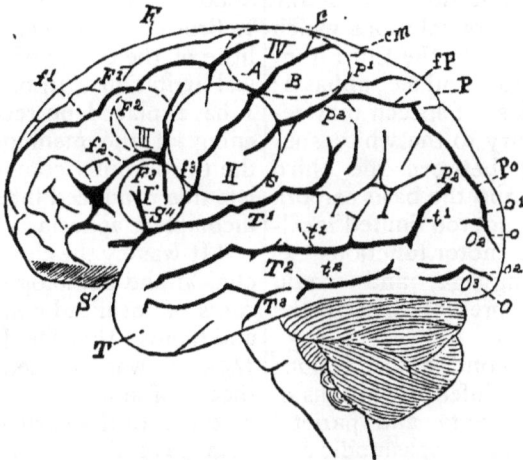

CORTICAL CENTRES OF THE HUMAN BRAIN.

S, Fissure of Silvius; c, Fissure of Rolando; po, Parieto-occipital fissure. A, Ascending frontal gyrus; B, Ascending parietal gyrus; F₃, Third frontal gyrus; P₂,′ Gyrus angularis. Circle I, Seat of lesions which (on the left side) cause aphasia. Circle II, Seat of lesions which convulse or paralyze the upper extremity of the opposite side. Dotted Circle III, Seat of lesions which probably convulse or paralyze the face on the opposite side. Dotted Oval IV, Seat of lesions which probably convulse or paralyze the lower extremity of the opposite side. These districts receive their blood supply chiefly from the middle cerebral artery. The remaining letters refer to anatomical points which explain themselves.

principal cerebral cortical centres, modified from FERRIER and ECKER, by SEGUIN, will be found very useful, as it embodies the results of the recent researches of FRITSCH and HITSIG, FERRIER, DALTON and SEGUIN;* particularly as this subject is now attracting much attention.]

The following is the summary given by SEGUIN (*loc. cit.*):—

PHYSIOLOGICAL.

"In the first place, it appears almost absolutely certain that in man a lesion involving the posterior part of the third frontal convolution (on the left side usually) causes aphasia; *i. e.*, impairment or loss of articulate speech, or even of language in general. It would seem, besides, that (1) lesions of the same part on either side of the brain produce paresis of many muscles concerned in lingual and pharyngeal movements; (2) that lesions of the anterior folds of the island of Reil (convolutions which are continuous with the third frontal), may also produce aphasia; and that (3) loss of speech may result from injury to the white substance lying between the third frontal gyrus and the basis cerebri. I believe in a not too limited localization of the motor functions exerted in language, and would graphically represent this by the circle marked I.

"In the second place, lesions limited to the inferior portions of the ascending frontal and parietal gyri have produced spasmodic and paralytic phenomena limited to the upper extremity of the opposite

PATHOLOGICAL.

"1. The symptoms of an *irritative lesion* of these parts consist in convulsions, with or without subsequent transient paralysis; *e. g.*, such a lesion in circle III would give rise to spasmodic movements in the superficial muscles of the face on the opposite side, with slight paralysis. Irritative lesions of the regions inclosed in circles II and IV will cause convulsions limited to, or first appearing in the hand and arm, or foot and leg, of the opposite sides. As regards circle I (Broca's speech centre), we know little of the effects of its pathological irritation. In one case which I have placed on record, a thickening of the meninges involving the third frontal convolution of the left side produced intermittent and incomplete aphasia.

"It was by the close study of the clinical and pathological aspects of cases of localized epilepsy (fingers and hands), that Dr. J. HUGHLINGS JACKSON was enabled to form his theory of motorial discharges from irritation of the cortex cerebri, and thus pave the way for FERRIER'S admirable researches. Dr. JACKSON must, I think, be considered, after

* See Lectures in New York *Medical Record*, delivered at the College of Physicians and Surgeons, New York, in January, 1878.

PHYSIOLOGICAL.

side. I am disposed to admit as highly probable that these parts are connected in the healthy living man with the various voluntary movements of the arm and hand. This zone is represented by circle II.

" I am not prepared to go further in admitting pathologically proved cortical centres, but would add that there are some reasons for believing that future autopsies will locate one centre for the external facial muscles just forward of the two centres named above, viz., the region included in the dotted circle III; and another for movements of the legs upon the upper parts of the ascending frontal and parietal, as roughly indicated by dotted oval IV."

PATHOLOGICAL.

Prof. BROCA, as the founder of our present growing doctrine of cortical localizations.

" 2. *Destructive lesions* of portions of the excitable district produce paralysis in peripheral parts across the median line. The symptoms will, to a certain extent, correspond with the *precise* location of the lesions, very much as in irritative lesions; *e.g.*, embolism of the first branch of the middle cerebral artery on the left side will cause softening of the posterior part of the third frontal gyrus, with the symptom aphasia. A destructive lesion of the principal part of the motor zone on the right side will produce left hemiplegia without aphasia; but if this lesion occupy the left hemisphere, loss of speech will co-exist with the paralysis."

It must be added that secondary descending degeneration ensues after destructive lesions of the motor regions of the cortex, and that we have late contracture or rigidity of the paralyzed limbs as part of the symptom group.

Negative characters of these cortical lesions are preservation of sensibility in the paralyzed parts, and (except with epileptic attacks) preservation of consciousness, and incompleteness of paralysis.

In diffused lesions of the cortex the chief symptoms are delirium, convulsions and pain; evidences of intense irritation. The coma and paralysis which follow may in some degree be caused by impaired nutrition of the cortex, but more probably by circulatory and tension changes in the whole encephalic mass.

As regards sensory cortical centres, Dr. SEGUIN believes that we have as yet no pathological data for their study.

7

DISEASE OF BRAIN CENTRES OTHER THAN CORTICAL.

The following tabular view of the paralyses, with the localization of the lesion, is mainly that of Professor DaCosta.*

SYMPTOMS.	SEAT OF LESION.
Hemiplegia, without disturbance of sensation. Incomplete paralysis of face. Electro-muscular contractility and tendon-reflexes normal or increased. Generally accompanied by apoplectic symptoms. Right-sided palsy usually accompanied by aphasia.	In corpus striatum, less markedly optic thalamus; on side opposite to hemiplegia.
Crossed paralysis (*i. e.* face of right and hemiplegia of left, or *vice versa*). Paralysis of face marked, one-sided loss of motion and sensation. General symptoms : giddiness, nausea. Albumen or sugar in urine.	Pons Varolii upon opposite side to palsy of limbs (below decussation of facial nerve).
Same as above, except complete facial paralysis (both sides of face).	Pons at level of decussation of facial nerve.
Paralysis of arm and leg, slight paralysis of face, dilatation of pupil of opposite side, with external squint (3d nerve paralysis).	Crus cerebri on side corresponding to affected eye. •
Paralysis of motion of arm or leg incomplete and transitory, soon followed by rigidity, no loss ·of sensation. Reflexes superfic al .and deep, preserved or increased.	Cortical part of brain in motor zone, on side opposite to palsy.
Paralysis of one arm and same side of face, sensation unimpaired; if palsy right-sided, aphasia.	Middle or lower third of the ascending convolutions in facial and manual centres, on side opposite to palsy.
Motion more or less completely affected on both sides of the body; sensibility diminished or lost on one side, increased on the other; the same as to temperature.	Medulla oblongata on side of increased sensibility and temperature, and at level of decussation of anterior pyramids.

* "Medical Diagnosis," 6th Edition, Philadelphia, 1884, page 122.

The observations of Brown-Séquard have demonstrated that in exceptional cases of brain tumor or lesion the symptoms do not correspond as accurately with the anatomical position of the lesion as is above indicated. At the present time these cases must be looked upon as really exceptional, and as not affecting the rules which have just been cited. More particularly are these aberrant symptoms likely to appear in tubercular disease of the brain. Indeed, Prof. Henoch (in *Charité Annalen*, fourth year), reports nine cases of tuberculosis of the brain that show how risky it is to localize, basing this upon recent physiological investigations. His results were as follows:—

SYMPTOMS.	LESION.
Case I.—Left hemiplegia.	Multiple tubercles of the cortical layer of both hemispheres, the frontal lobes and tubercle of the left half of the cerebellum.
Case II.—Tremor and paresis of the right side, finally, contraction of all extremities.	Tubercle of the left frontal lobe, the left corpus striatum, both thalami and right half of the cerebellum.
Case III.—Hemiplegia and contracture of the left side, as well as of the facial nerve.	Tuberculosis of the right frontal lobe.
Case IV.—Contracture and involuntary motion on right half of face and body.	Tuberculosis of the left frontal lobe.
Case V.—Complete absence of symptoms until meningitis set in.	Tuberculosis of the commissure of the cerebellum and of both hemispheres.
Case VI.—Paralysis of the left abducens, the left iris and right arm.	Tuberculosis of the commissure of the cerebellum.
Case VII.—Absent, until meningitis sets in.	Tubercle in the pons.
Case VIII.—Completely absent.	Tubercle of the left posterior lobe.
Case IX.—Paralysis of the right abducens.	Tuberculosis of both posterior lobes, the posterior corpora quadrigemina, the pons and left crus cerebelli.

L. c f C.

Of all these cases only II and III show the possibility that lesions of the motor centres of the frontal convolutions produce motor lesions of the opposite side. This chance of diagnosis, however, is very limited, as is shown by the other cases where these locations were free from disease, and yet the same symptoms produced with lesions in other parts of the brain, even cerebellum (Case VI). Sometimes the intensity of the symptoms does not seem to correspond with the intensity of the lesion (V and VI). HENOCH believes that a close study of the fibres leading from and to these physiological centres will do much to reconcile the apparent contradictions between pathological and symptomatological differences.*

(The subject of insanity will be separately considered at the end of this section.)

INFANTILE CEREBRAL PARALYSIS.

Dr. R. NORRIS WOLFENDEN, in the *Practitioner* for September, 1886, describes four cases of this affection. The disease has also been called spastic cerebral hemiplegia, and, again, *poliencephalitis acuta*. These names, among them, manage to describe something of the disease. The last, as indicating the pathology, and being analogous to *poliomyelitis*, is on some accounts the best, were it not too technical and pedantic. One of these cases was briefly as follows: A boy of ten months was attacked with a series of convulsions, followed by paralysis of the left arm and leg. For six months the leg and arm were spastic; then they relaxed, leaving permanent weakness. Epileptiform attacks continue at intervals of about twice a week. Intelligence had been much impaired, so that at eight years he could not be taught to read. Weakness of the arm and leg continued, and the fingers were subject to slight athetotic movements. The tendon-reflexes of this side were exaggerated. There were no reactions of degeneration.

The disease is distinguished from *spinal* paralysis by the affections of the intellect, speech, special senses (sometimes), and the absence of impaired sensibility. In the spinal form also there is atrophy and coldness, with progressive reactions of degeneration, while the spastic condition is not an early and characteristic sign. Club-foot deform-

* Cincinnati *Lancet and Clinic*, May 31, 1878.

ities are more constant and persistent in the latter. The cerebral form may be said to occur more as a hemiplegic and epileptic type. The disease has followed dentition and the eruptive fevers, but in many cases the causes cannot be determined. It is an affection of early infancy.*

SPINAL DISEASES.

A leading symptom of many diseases of the spinal cord, whether functional or organic, is *paraplegia*. This is so rarely of cerebral origin that ordinarily the brain may be omitted from the discussion, unless there is the coëxistence of distinct evidences of brain disease, as headache, impaired cerebration, affections of special senses, and paralysis of parts supplied by nerves arising above the spinal cord.

The following classification of diseases giving rise to paraplegia, with their characters, has been proposed by Prof. H. C. Wood.†

ORGANIC.	FUNCTIONAL.	HYSTERICAL.
Disease of the cord.	Anæmic.	Hysteria.
	Reflex (from peripheral irritation, renal, preputial, etc.)	
	Dyscrasic (diphtheritic, etc.).	

The last-mentioned, hysterical, is also functional, but simulates the organic more closely than does the second group. (For further consideration of Hysterical Paralysis, see Hysteria.) It must be admitted, however, that so-called functional disorder cannot long exist without being followed by change in structure.

The general distinctions between the organic and functional paraplegias may be presented as follows:

ORGANIC.	FUNCTIONAL.
Onset may be almost instantaneous or very rapid, though sometimes gradual.	The onset always more or less gradual, except in the hysterical form, where the paralysis may be abrupt.
Usually at some period spasm or pain in the affected limbs.	Spasms or pain rarely or never present.

* J. H. Lloyd, *Phila. Med. Times*, January 22, 1887.

† "On the Diagnosis of Diseases Accompanied by Paraplegia." 1875.

ORGANIC.	FUNCTIONAL.
Often a sensation of a band or stricture around the waist, *girdle-pain* (pathognomonic).	Not found.
Anæsthesia frequent and often complete.	Anæsthesia not observed, or but slight.
Retardation of sensation (a perceptible time elapses between the patient's seeing his feet touched and feeling that they are) (pathognomonic).	Sensation, if present at all, is not retarded.
Symptoms of paralysis of the bladder.	No symptoms whatever of vesical paralysis, except in the hysterical form.

Where the bony canal is involved and caries is present, this condition may generally be discovered by ROSENTHAL'S test. This consists in passing down the back a pair of electrodes attached to a faradic battery of some power, one pole being placed upon each side of the spine. Under these circumstances if there be any caries or inflammation of the vertebræ, the moment its locality is reached, the patient starts or screams, from the burning, sticking pain caused by the passage of the galvanic current through the inflamed tissue. Dr. WOOD states that he has not found this test as trustworthy as its originator claimed it to be, and as, apparently, it ought to be. In cases simulating caries, however, the pain is probably not so severe as where the vertebræ are really affected. Moreover, absence of the pain in any case seems to be conclusive evidence of the non-existence of bone disease.

The following study of the principal organic spinal diseases, from the writings of SEGUIN, CHARCOT, and other authorities, when taken in conjunction with the tabular view of paralysis, will often enable the diagnostician to determine both the nature and location of a spinal lesion:

DISEASES OF SPINAL CORD.

Transverse diffused myelitis (acute and chronic) { occupying the entire section of a limited portion of the cord, more or less completely.

Disseminated sclerosis (*sclérose en plâques*) { Patches of disease situated primarily in the connective tissue, and scattered without regard to the "systematic" grouping of the nervous elements.

Degenerative disorders, mainly affecting the columns of the cord.	Sclerosis of the posterior columns. (Locomotor ataxia). Duchenne's Disease. — Its distribution is "systematic," and *probably*, it is essentially a primary disease of the nerve elements rather than of the connective tissue.
	Symmetrical lateral sclerosis. (Paralysis spinalis spastica.) — Ditto, though its pathology is as yet almost purely a matter of inference. Its characteristic symptom is muscular rigidity.
Myelitis of the gray matter of the anterior cornua.	Poliomyelitis anterior. — Acute. — Infantile paralysis. Acute spinal paralysis of the adult. Subacute. Chronic.
	Progressive muscular atrophy and progressive bulbar paralysis ("labio-glosso-pharyngeal paralysis"). — Often classified as a special form of poliomyelitis chronica, but characterized by the absence of paralysis, except such as is directly due to the muscular atrophy.
Antero-lateral sclerosis (amyotrophic).	Not yet thoroughly studied, but believed by CHARCOT and others to involve at once the lateral columns and the anterior cornua; the characteristic symptoms being atrophy with contracture, beginning in the upper extremities.

TABULAR VIEW OF SPINAL PARALYSIS.

SYMPTOMS.	SEAT OF LESION.
Paralysis of compressor urethræ, accelerator urinæ and sphincter ani. No paralysis of muscles of the legs.	In the termination of the cord, low down in the sacral canal.
Paralysis of muscles of bladder, rectum and anus. Loss of sensation and motion in muscles of legs, except those supplied by anterior crural and obturator nerve.	In the cord, at the upper limit of the sacral region.
Both legs paralyzed as to sensation and motion. Loss of power over bladder and rectum. Lateral muscular walls of abdomen paralyzed, thus interfering with expiratory movements of respiration. Electro-muscular contractility diminished or lost.	In the cord, at the upper limit of the lumbar region.
Paralysis of legs, etc., as above. Paralysis of all the intercostal muscles, and consequent interference with inspiration. Paralysis of muscles of upper extremities, except those of the shoulders, which receive their nerves from the higher portions of the cervical region.	In the cord, low down in the cervical region.

TABULAR VIEW OF SPINAL PARALYSIS.

SYMPTOMS.	SEAT OF LESION.
In addition to the preceding, diffculty of swallowing and vocalization, contraction of pupils, palpitation of heart and priapism.	In the cord below the middle cervical region.
In addition to above, paralysis of the phrenic nerve and diaphragm, of the scaleni, intercostales, serrati magni, and many of the accessory respiratory muscles which act upon and from the shoulder. Death resulting at once from suspension of all respiratory movements.	In the cord, at or above the middle of the cervical region, or the level of the fourth cervical pair of spinal nerves.
Paraplegia developing itself symmetrically.	Anterior half of the medulla spinalis or its sheaths.
Paraplegia of the legs.	Dorso-lumbar enlargement of cord.
Paraplegia of the arms.	Cervical enlargement of cord.
Cerebral paraplegiæ, so-called, are very rare, and are in reality two distinct hemiplegiæ.	In both sides of the brain. Exceptions in cases of disease of the medulla oblongata (very rare).
Paraplegia from disease of the vertebral column.	Roots of spinal nerves at point of injury, especially posterior roots, which long remain in a state of painful excitation.
Characteristic symptoms of tabes dorsalis or locomotor ataxia.	Posterior part of med. spinalis.
Progressive muscular atrophy.	Gray substance of spinal cord vicinity of the central canal or diffused through anterior roots.
Hemiplegia with crossed hemianæsthesia.	In one lateral half of spinal cord. The hyperæsthesia of the paralyzed side is probably due to paralysis of the vaso-motor nerves of that side.

TABULAR VIEW OF SPINAL PARALYSIS.

SYMPTOMS.	SEAT OF LESION.
· Bilateral neuralgia of the legs and arms accompanying symptoms of tabes dorsalis.	In posterior roots of spinal nerves and their prolongation into the gray substance of the cord.
Bilateral contractions affecting the extensor muscles.	In the spinal cord.
Unilateral contractions affecting the flexor muscles.	In the brain.

Diseases of the spinal marrow have been classified by Dr. WOOD according to the *rapidity of their onset*, as follows, the attack being considered rapid when decided paralysis has developed within forty-eight hours;

RAPID ONSET.	SLOW ONSET.
Congestion.	Sexual exhaustion.
Meningeal apoplexy.	White softening.
Spinal apoplexy.	Chronic myelitis.
Acute myelitis.	Tumors.

In *congestion of the cord* the diagnosis rests upon: Suddenness of onset; uniform, bilateral loss of voluntary motion, reflex activity and sensation; absence of all symptoms of irritation, such as spasms or violent pains; absence of constitutional disturbance. It must also be remembered that the palsy affects first and most severely the lower limbs, but may rise to the arms, and, finally, to the muscles of respiration, and thus prove fatal; that so far as the paralysis extends, all the muscles are involved; that motion is affected more than sensation; and that very rarely, if ever, does ulceration or other indication of trophic changes occur.

In *meningeal spinal apoplexy* the symptoms are also due to pressure, but the effused blood not only disturbs the cord by pressing upon it, but also irritates the membranes and the nerve-roots, especially when first thrown out. Consequently, in the first few hours or days of a meningeal hemorrhage, there are violent spasms and pains, due either to an incipient meningitis, or more probably to a direct irritation of the nerve-roots. The extent and amount of the symptoms vary, of course, with the position and amount of the hemorrhage. Later there are symptoms of pressure, varying in intensity with the amount of the effusion ; and absence of febrile symptoms, unless decided meningitis be produced by the clot.

In true *spinal apoplexy* the symptoms come on with absolute abruptness. The cord is so small a body that a clot in its substance interrupts at once its function. The paralyses of motion and sensation are complete, and reflex movements are greatly exaggerated. As there is no correlation of the spinal nerve-roots, the spasms and pains of meningeal hemorrhage are wanting.

Acute primary myelitis is a very rare affection. The diagnosis should present no difficulty. The distinct febrile reaction, which is stated to be always present, separates it at once from all other acute affections of the cord proper, so that it can be confounded only with acute meningitis. Probably, in the majority of cases, it exists coincidently with this disorder; but even when it is isolated, the symptoms at first closely simulate those of meningitis.

COMPARISON OF ACUTE SPINAL DISEASES.

MYELITIS.	MENINGITIS.	CONGESTION.
Constant pain in the spine at a point corresponding with the upper limit of inflammation, rendered more acute by pressure on vertebral spine.	Pain usually rheumatic in character, diffused along the spine, not increased by pressure ; but augmented by flexions of trunk.	Formication alternating with numbness in the beginning of the attack, especially in fingers and toes.
The alternate application of ice and hot sponge to spine causes the same burning sensation at seat of lesion, but above it the sensation is normal.	Nerves coming out through the inflamed part of the meninges, the seat of acute pain, much increased by movements of limb.	Only slight pain in spine, scarcely increased by pressure.
Sensation as of a cord or ligature around the body at the limit of paralysis always present when dorsal region is affected ; when higher up spasm of the sphincters and priapism often occur.	Frequent spasms of muscles of the back. Spasm of sphincter vesicæ may occur, followed by retention of urine and paralysis. Convulsive movements of paralyzed parts.	Frequently hyperæsthesia ; sphincters more paralyzed than in other forms of paralysis (BROWN-SÉQUARD).

COMPARISON OF ACUTE SPINAL DISEASES.

MYELITIS.	MENINGITIS.	CONGESTION.
Paraplegia complete.	Paraplegia varies in degree, sometimes increasing and subsequently rapidly diminishing.	Paralysis generally not limited to lower limbs, but involves upper extremities and respiratory muscles. In some cases power of moving paralyzed legs is better after resting; ordinarily, however, the paralysis is worse on first rising in the morning.
Anæsthesia or paræsthesia (except when gray matter is not involved, which is rare), muscular sensibility much impaired, early.	Anæsthesia very rare; generally hyperæsthesia.	Frequently morbid increase of sensibility.
When disease is high up in dorsal region, energetic reflex movements may be produced.	Increased reflex movements, which cause pain, may be excited.	Slight spasmodic movements sometimes observed in paralyzed parts.
Marked tendency to bed sores; sloughs form early on sacrum and nates.	Less marked in uncomplicated cases of meningitis.	Ulceration occasionally happens.

CHRONIC SPINAL DISORDERS.

In the slow or chronic forms of spinal disease, *spinal tumors* may be considered first. There are three classes of phenomena to be looked for in this disease: local symptoms of diseased structures; atrocious pains at a distance from the seat of the disease, due to the involvement of nerve-roots and nerves, where they pass through the inflamed tissues; and paralytic symptoms, the results of pressure, and to some extent of a local myelitis. In cases of suspected tumors of the spine all these symptoms are to be sought after. In cancer they are often all present, and the distant pains are especially remarkable for their atrocity.

The other chronic spinal diseases may be classified with reference to the characteristic of *tremors* as follows:

WITHOUT TREMORS.	WITH TREMORS.
Sexual exhaustion.	Paralysis agitans.
White softening.	Multiple sclerosis.
Chronic myelitis { softening. { sclerotic.	
Local myelitis.	

The difference between *sexual exhaustion* and *myelitis* is probably one of degree only; but the former is curable, the latter is not.

CHRONIC DEGENERATIVE DISEASES OF THE CORD.

In distinguishing the various forms of *disseminated* or *multilocular* cerebro spinal affections, the following table, given by Professor CHARCOT, will render valuable assistance. The symptoms of greatest importance are in *italics*.

CEREBRO–SPINAL AFFECTIONS.

	LOCOMOTOR ATAXIA.	MULTILOCULAR SCLEROSIS.	DISSEMINATED SYPHILOSIS.	GENERAL PARALYSIS.
CEPHALIC SYMPTOMS.	Epileptiform Apoplectic Attacks.	*Epileptiform Apoplectic Attacks.*	Epileptiform Attacks. Paraplegic Hemiplegic Epilepsy.	Epileptiform Apoplectic Attacks.
	Vertigo. Diplopia, Strabismus.	*Vertigo.* Diplopia.	*Vertigo.* Diplopia.	*Vertigo.* Diplopia.
	Amaurosis.	*Nystagmus.* Amblyopia, White Atrophy.	*Amblyopia, Optic Neuritis.*	Amblyopia.
	Inequality of Pupils. *Facial Anæsthesia.*		*Headache, Fixed Pain.*	*Inequality of Pupils.* Headache.
	Deafness. *Ménière's Vertigo.* Embarrassment of Speech. Laryngismus.	*Embarrassment of Speech.* *Difficult Deglutition.*		*Embarrassment of Speech.*
		Pneumogastric Palsy.	Total Facial Palsy.	
VISCERAL SYMPTOMS.	*Gastric Crises.* *Nephritic Crises.* *Vesical Crises.* *Paresis of Bladder.* *Cystitis.*	*Gastric Crises.*	Non-nervous Crises.	Paresis of Bladder.
SPINAL SYMPTOMS.	*Girdle-pain.* Hyperæsthesia, Anæsthesia. *Incoördinated Movement.* Contractures and Trepidations.	Lightning pains. *Plaques.* Incoördination. *Special Trembling.* Spasmodic Paraplegia.	*Pseudoneural Pains.* *Spinal Hemianæsthesia.* *Spasmodic Paraplegia under form of Hemiparaplegia.*	*Lightning Pains.* Tingling. *Incoördination.* *Special Trembling of Hand.* Paresis. Trepidation.
TROPHIC SYMPTOMS.	Eschars. Arthropathies. Fractures. Muscular Atrophy.	Eschars. Arthropathies. Muscular Atrophy.		Eschars. Muscular Atrophy.

In applying these symptoms in practice, we should, of course, give first attention to those which are most characteristic. Thus, if we observe, in a patient, ataxy with nystagmus, we think at once of multilocular sclerosis and not of locomotor ataxy (tabetic series), because nystagmus is a valuable symptom of multilocular sclerosis. In the same way spasmodic paraplegia (recognized by the continual trembling movements which are produced when a single blow is struck upon the muscle) we find is produced by a localized lesion in the cord, more particularly involving the lateral columns.

In order that these forms shall be better understood, it may not be out of place to review some of the chief clinical manifestations of sclerosis of the cord.

In *sclerosis of the antero-lateral white columns*, Dr. GOWERS* states that there is loss of voluntary power below the lesion, descending degeneration in the anterior and lateral columns (direct and crossed pyramidal tracts, especially the latter), and over-action of the lower centres. This over-action may be manifested only as excessive knee-reflex† and developed ankle-clonus (tendon-reflex), or it may increase from this to spasm and rigidity—spastic paraplegia. There is no wasting unless the degeneration extends from the lateral columns to the anterior cornua. Then we have a combination of spasm and wasting, in which, if the cornual degeneration proceeds, the spasm and rigidity may lessen as the degeneration advances. In disease limited to the lateral columns (at any rate, when the disease is limited to the pyramidal tracts) there is no loss of sensation or incoördination, and no interference with the nutrition of the skin. These symptoms of "spastic paraplegia" may arise from a primary degeneration in the lateral columns, limited thereto (*lateral sclerosis*). Such cases are extremely rare, and in the majority the disease is a focal lesion more or less extensive at some level in the dorsal or cervical cord, and the degeneration in the lateral columns is secondary. The evidence of the latter form is afforded by the frequently sudden or rapid onset of the symptoms in the first instance (primary sclerosis being always gradual in onset), and the evidence which may generally be discovered that there

---

* Address delivered before the Medical Society of Wolverhampton, Oct. 7th, 1879.

† See page 102, note on Patellar-Tendon Reflex.

has been at some time, or is in some region, a lesion, which extends beyond the lateral columns. Descending lateral sclerosis, with secondary spasmodic phenomena in the limbs, may even result from damage to the motor tracts above their decussation—in the medulla, the pons, or the motor parts of the cerebral hemispheres. It occasionally results from bilateral injury to the surface of the brain during difficult birth, but such cases are very rare.

2. In *disease of the posterior columns* there is interference with coördination without loss of power; eccentric pains, impaired sensation and diminution of reflex action, in consequence of the implication of the sensory roots. All these symptoms depend on disease of the root-zone of the posterior columns. Disease of the posterior median column gives rise to no known symptoms.

The posterior columns may be damaged by any pathological process, and they are frequent seats of primary degeneration. The symptoms of locomotor ataxy usually present the following order Loss of the deep reflexes, pains, incoördination, diminution of sensation, loss of the superficial reflexes, occasionally interference with the nutrition of bones and joints.

There is no loss of motor power or wasting as long as the disease remains limited to the posterior columns. It may, however, extend forward into the anterior cornua, causing muscular atrophy and weakness to be conjoined with the ataxy. Or the lateral columns may be affected at the same time as the posterior; we then have weakness as well as ataxy, but no wasting. The disease of the lateral columns causes increase of the deep reflexes, and this increase may thus coëxist with incoördination, the increased action of the reflex centres being so great that they are not arrested by the damage to the posterior root (which is often, in these cases, slight). Thus we have the anomaly of ataxy with excess of the tendon reflex instead of its loss.

An important fact to remember regarding the posterior columns is their proneness to degenerate; they recover less readily than any other part of the cord. A lesion in one spot may set up a degeneration which ultimately involves them in their whole extent. Damage affecting the whole thickness of the cord may pass away from the rest and persist in the posterior columns, and even spread there. In such a case we have ataxy succeeding loss of power.   Strength returns, incoördination remains.

3. The anterior cornua contain the motor nerve-cells, which (1) influence the nutrition of the motor nerve fibres proceeding from them, and consequently that of the muscles; (2) constitute the terminal link in the path of the voluntary impulse from the brain to the muscles; (3) form part of the reflex loop, probably also of the reflex centre, to which these muscles are connected.

• Hence we have as the result of disease of the anterior cornua, (1) degeneration of the motor nerves and wasting of the muscles; (2) loss of voluntary power, *i. e.*, paralysis of those muscles; (3) interference with or arrest of the reflex actions in which these muscles take part.

The extent of these symptoms, whether they are unilateral or bilateral, affect many muscles or few, will depend strictly on the extent of the disease in the spinal cord.

Of the three symptoms, the muscular wasting is incomparably the most important.   Paralysis may result from disease elsewhere in the motor tract, *i. e.*, disease of the lateral column higher up.   Loss of reflex action may depend on disease elsewhere in the reflex loop, *i. e.*, disease of the sensory fibres in or outside the cord.   But muscular wasting is due only to a lesion of the motor cells, or to a lesion of the nerves, cutting the muscles off from the influence of these cells.   In most cases we are able to exclude the latter without difficulty; the state of muscular nutrition comes thus to be of the highest importance as indicative of the state of the anterior cornua of the cord.

Disease of the anterior cornua is often combined with disease of the lateral (pyramidal) columns similar to the descending degeneration. CHARCOT believes that in these cases of degeneration in the lateral column is primary, its symptom, muscular rigidity, preceding the symptom of the cornual disease, muscular wasting, and he terms the affection " lateral amyotrophic sclerosis."   GOWERS believes, however, that this position needs reconsideration, and that the degeneration in the lateral columns is, sometimes at least, secondary to, or simultaneous with, the disease in the cornua.   It often spreads, however, beyond the fibres related to the degenerated cornua, and so may cause weakness and spasm in the limbs below the seat of the muscular atrophy. Thus we have wasting in the arms, and weakness with spasm in the legs, and even, as I have seen, wasting in the shoulder-muscles, and weakness without wasting in the hands.

Certain lesions may damage the motor tracts slightly and impair conduction in a peculiar way, rendering it apparently unequal in different fibres. As a consequence, the muscular action is unequal in different muscles, and intead of a balanced coördinated movement, we have an unbalanced jerky movement. This is seen especially when irregular islets of sclerosis affect the cord—disseminated or insular sclerosis—and, according to the researches of CHARCOT, it appears that this irregular conduction is the result of the unequal wasting of the medullary sheaths, the axis-cylinders remaining. A precisely similar symptom may result from pressure on the motor tract—as by a growth. Not rarely this "disseminated" or "insular" sclerosis in one region is combined with a system-degeneration in another. An occasional combination, for instance, is the jerking movement (from cervical insular sclerosis) in the arms, and weakness with spasm (from lumbar lateral sclerosis) in the legs.

4. A total transverse lesion of the cord at any level, however limited in vertical extent, separates all parts below the lesion from the brain, and hence, so far as will and perception are concerned, produces the same effect as if the whole of the cord below the lesion were destroyed. A section across the cord in the middle of the cervical enlargement, for instance, paralyzes all parts below the neck. Hence the extent of the paralysis indicates only the upward extent of the lesion. This is also indicated by the position of the girdle pain, or zone of hyperæsthesia, which is due to the irritation of the sensory roots in the lowest part of the upper segment—an important indication when the lesion is in the dorsal region, where the precise limitation of motor weakness may be recognized with difficulty.

The Tendon Reflex.—For the *diagnosis* of posterior sclerosis, WESTPHAL has noted the following symptom: "If a healthy man sits with one knee-joint resting upon the other (a very common attitude), and the ligamentum patellæ of the supported leg be smartly struck just below the knee-cap with the side of the hand, a sudden contraction takes place in the quadriceps femoris muscle (of which the ligamentum

patellæ represents the tendon), and the foot is consequently jerked upward to a degree which varies in different individuals. Now, in confirmed examples of locomotor ataxia this reaction does not take place.
No matter on what part of the ligament below the knee-cap, or with
what force the blow is struck, the foot hangs motionless. In order to
establish with accuracy the absence of the phenomenon, certain precautions ought to be taken. The leg should be bare ; the patient must
not offer voluntary resistance to the movement of his leg, and the ligament should be struck with some hard implement which can be
swung like a hammer. An ordinary wooden stethoscope answers very
well if it is held loosely by the small end, and the blow given with the
edge of the ear-piece. But, however administered, several blows
should be struck on the ligament, slightly changing the position each
time, as there is generally one spot from which the reaction is peculiarly energetic. This is usually a little below but very near to the
patella. Ankle-clonus may be similarly developed by tapping the
tendo-Achillis.

The following are the conclusions given by ERB* in regard to the
interpretation of symptoms:

In diseases of the spinal cord, paralysis rapidly followed by a
marked degree of atrophy and by the reaction characteristic of degeneration, points to disease of the anterior roots (rarely), or of the gray
anterior cornua (more frequently). In this case all reflex actions are
absent.

Paralysis with tension and contraction of muscles, without atrophy,
is very probably due to some affection of the lateral columns.

Paralysis without loss of reflex function and without atrophy, points
to an affection of the parts which ascend to the brain, outside of the
gray substance, or, at least, outside of the ganglia of the anterior cornua. Such are mostly cases of circumscribed disturbances of conduction, the end of the cord below the lesion remaining intact.

Paralysis, with trophic disturbances, gives room for suspecting an
affection of the gray substance, since primary affections of the roots
are rare.

Very extensive palsy, with much atrophy, the reaction of degenera-

* From a review in *Journal of Nervous and Mental Diseases*, Chicago, Oct., 1878.

8

tion, absence of reflex acts, points to a widely diffused lesion of the anterior gray substance.

Paralysis in the districts supplied by certain pairs of roots (both arms alone, or both crural nerves) points to a strictly localized affection of roots, or lesion of the gray anterior cornua. The conclusions in regard to the *nature* of the lesion in the cord are far less certain than those relating to its place.

Cases of spinal paralysis, accompanied by atrophy of the muscles, whether in children or adults, acute or chronic, are described under the heads of *poliomyelitis anterior, acute, and chronic.*

Destruction of the central trophic apparatus, or its separation from the peripheral parts, produces the symptoms of degenerative atrophy. "Upon the whole, we are justified in assuming a disease of the anterior cornua when the electrical examination shows the existence of the reaction of degeneration, and consequently of degenerative atrophy of nerves and muscles, *provided the disease is clearly of spinal origin.*" (ERB).

In infantile palsy (lesion in the anterior cornua), observers are not agreed as to whether the change in the ganglion cell is primary, or whether it is the consequence of an interstitial myelitis.

The following table will be found valuable in diagnosticating certain chronic disorders.

CEREBRO-SPINAL SCLEROSIS.	PARALYSIS AGITANS.	LOCOMOTOR ATAXIA.
Disease of adult life.	In old persons chiefly.	In adults.
Tingling and numbness; diminished muscular power, chiefly in the legs.	Felt mainly in the arms.	Tingling and numbness of legs without loss of power (want of coördination exists).
Eye symptoms absent in spinal form; when they occur in cerebro-spinal form they are persistent and progressive.	No proper eye symptoms.	Ocular troubles, defective vision and accommodation, strabismus, ptosis or double vision. These symptoms temporary.
Tremor or trembling follows the paralysis.	Precedes paralysis.	Absent.
One or both limbs paretic, ultimately becoming completely powerless.	Muscular weakness in one or both arms, and then extends into lower extremities. Only rarely passing into true paralysis.	No paralysis during the early stages.

CEREBRO-SPINAL SCLEROSIS.	PARALYSIS AGITANS.	LOCOMOTOR ATAXIA.
In the paretic stage the gait is distinctive; the foot is swung around, describing an arc of a circle, and brought flatly upon the ground. With this eccentric curvilinear projection of the foot there is an exaggerated alternate semi rotation of both halves of the pelvis.	In attempting to walk, first balances on his feet, and starts with head and trunk bent forward on the toes or fore part of feet, and with short steps goes hopping and trotting along at almost running speed (festination).	A stumbling, staggering gait, without true paralysis.
No spontaneous tremor; always caused by motion or excitement.	Trembling early, incessant, even when at rest; scarcely interrupted by sleep.	No tremor.
Nystagmus, usually binocular.	Never met with.	Not present.
Articulation slow and scanning.	Articulation indistinct; embarrassed.	Not affected.
Intellect early impaired.	Unaffected until late.	Not marked.
Boring, gnawing, and lancinating pains rarely complained of.		Such pains frequently precede the loss of motion.
Early paresis, passing into paralysis, is characteristic.		Paraplegia always a late phenomenon.

(For diagnosis of Locomotor Ataxia from General Paralysis of the Insane, see page 123, under this head.)

Paralysis may also be caused by reflex irritation, and closely simulate organic disease of the cord. BROWN-SÉQUARD* gives the following points of distinction (with unimportant additions):

PARAPLEGIA.

FROM REFLEX IRRITATION.	FROM MYELITIS.
1. Is *preceded* by an affection of uterus, bladder, kidneys, or prostate gland. May be caused by phimosis.	1. Usually no disease of the genito-urinary organs except as consequent on the paralysis.
2. Usually lower limbs alone paralyzed.	2. Usually other parts paralyzed besides the lower limbs.
3. No gradual extension of the paralysis upward.	3. Most frequently a gradual extension of the paralysis upward.

* "Lectures on the Diagnosis and Treatment of Paraplegia," p. 33.

PARAPLEGIA (*Continued*).

FROM REFLEX IRRITATION.	FROM MYELITIS.
4. The paralysis is usually incomplete, an extreme debility or weakness of the limbs rather than paralysis.	4. Very frequently the paralysis is complete.
5. Some muscles more paralyzed than others.	5. The degree of paralysis the same in the various muscles of the lower limbs.
6. Reflex power neither much increased nor completely lost.	6. Reflex power often lost; or sometimes much increased.
7. Bladder and rectum rarely paralyzed, or at least only slightly so; sphincter ani weak.	7. Bladder and rectum usually completely paralyzed, or nearly so.
8. Spasms in paralyzed muscles extremely rare.	8. Always spasms, or, at least, twitchings.
9. Very rarely pains in the spine, either spontaneously or on application of pressure, percussion, or a hot, moist sponge, or ice.	9. Always some degree of pain existing spontaneously, or caused by external excitations.
10. No feeling of pain or constriction around the abdomen or chest.	10. Usually a feeling as if a cord were tied tightly around the body at the upper limit of the paralysis.
11. No formication, pricking, or disagreeable sensations of cold or heat.	11. Always formications, or pricking, or both, and very often sensations of pricking or heat or cold.
12. Anæsthesia rare, the tactile sensibility being but slightly, if at all, impaired; but the muscular sense is almost lost.	12. Anæsthesia very frequent and always at least numbness.
13. Usually obstinate gastric derangement.	13. Gastric digestion good, unless the myelitis has extended high up in cord.
14. Variations in the degree of the paralysis corresponding with changes in the primary disease.	14. Ameliorations very rare, and not following changes in condition of the urinary organs.

PARAPLEGIA (*Continued*).

FROM REFLEX IRRITATION.	FROM MYELITIS.
15. Usually the urine is acid, unless the urinary organs are diseased.	15. Urine almost always alkaline.
16. Cure of the paralysis frequently and rapidly obtained, or taking place spontaneously after a notable amelioration or cure of the genito-urinary affection.	16. Frequently a slow and gradual progress towards a fatal issue, and rarely a complete cure.
17. Usually muscles do not become atrophied, and temperature is little lowered.	17. Atrophy of muscles of the paralyzed parts.
18. Therapeutic results good.	18. Treatment of little benefit.

Mr. GOWERS divides spinal lesions, according to the time required in their development, into six classes, whose comparative features are shown in the following table:

Pressure or Growths.

Sudden (*few minutes*).
Acute (*few hours or days*). } Vascular lesions.
Sub-acute (one to four weeks).
Sub-chronic (one to two months). } Inflammation (myelitis).
Chronic (two to six months).
Very chronic (six months and upward). } Degeneration.

He recommends, in examining a case of disease of the spinal cord, to follow a definite plan. "First endeavor to ascertain the exact seat of the lesion; note how far the several conducting functions of the cord are impaired, and the highest level of their impairment; then ascertain the condition of the central functions, and especially muscular nutrition, and irritability and reflex action (first in the part below the level at which conduction is impaired and secondly, at the supposed level of the lesion; and in this way you may infer, without much difficulty, what is the extent of the lesion transversely and vertically). In the next place, endeavor to ascertain its nature by considering, first, how the symptoms came on and developed; secondly, which of the lesions having this onset and development are common in the region affected; and, thirdly, which of them are produced by the cause or causes to which the disease is apparently due."

Some special forms of paralysis require separate discussion.

PSEUDO-HYPERTROPHIC PARALYSIS.

This is a disease of children, usually attacking them after the second year of life. At this period it is found that when they are placed upon their feet they fall down, or clutch at the nearest object, to support themselves; or in other cases it may be that the child has commenced to walk, when, without pain or fever, or sometimes after convulsions, it is found to be soon fatigued, either by walking or standing, and at length it can no longer walk or hold itself upright; or, again, it may be that the child does not walk until very late—2½ or three years—and then very feebly and imperfectly.

In the advanced stage the child is unable to rise from the floor in the ordinary manner. He drags himself up with his hands; or if he be lying down and no support be forthcoming, he gets upon his hands and knees, and then, grasping each thigh alternately, is able to raise himself sufficiently from the floor to get first one and then the other foot upon its sole. He then lays hold of his thighs with successive grasps, one above the other, and thus, as it were, climbs up them to a standing position. This method of getting on the feet is pathognomonic of pseudo-hypertrophic paralysis.*

Symptoms.—The principal morbid phenomena are (DUCHENNE):

1st. In the beginning, feebleness of the lower limbs.

2d. Lateral balancings of the trunk and widening of the legs during walking. Inability to raise up on the toes when standing.

3d. A peculiar curvature of the spine (ensellure), or saddle-back (lordosis), both in walking and standing.

4th. Equinism (talipes equinus), with a peculiar over-extension of the first phalanges of the toes, which DUCHENNE calls "griffe des orteils."

5th. Apparent muscular hypertrophy.

6th. Stationary condition.

---

* Nervous Diseases and their Diagnosis. By II. C. Wood, M. D., Phila., 1887, p. 84.

7th. Generalization and aggravation of the paralysis.

8th. Patellar-tendon reflex abolished in advanced stage. No paralysis of the bladder or rectum.

When the disease has arrived at the stage of apparent hypertrophy, the appearance of the patient is very characteristic, and its true nature would be at once obvious to any one who had any knowledge of its symptoms; but in the earlier stages there is but little to guide us to a diagnosis unless we have some hereditary history. Of the hereditary nature of this affection the published cases give ample proof.

There are apparently two forms of this disease—one of spinal, the other of muscular origin.

PARALYSIS, FROM LEAD POISONING AND HYSTERIA.

In this form of paralysis the usual diagnostic symptoms, to wit, a history of exposure to lead, the blue line on the gums, constipation, and colic, may all be absent; hence the diagnosis must rest upon the peculiar characters of the palsy—especially the effects of electric currents upon the muscles. These are the only reliable evidences of the nature of the disease. These characteristic reactions, first described by DUCHENNE, are as follows:

Excitability to faradaism absent or sensibly diminished in all the muscles of the forearm except the supinators longus and brevis. In health the supinator brevis cannot be directly faradized, on account of its deep position. But in lead palsy it very often happens that the wasting of the extensor communis digitorum has proceeded far enough to uncover the supinator brevis sufficiently to allow a small rheophore to be applied to it in the space of about a square inch at the upper and back part of the forearm. If it be found (both arms being affected) that the common extensor fails to respond to faradaism while the short supinator close by, on a lower plane, is readily excited by it, the case may be positively set down as one of lead palsy.

Hysterical Paralysis, in spite of its frequent close imitation of the organic forms, is readily diagnosed by attention to the following points:

1. In hysterical hemiparesis the face is rarely, and the tongue never, affected.

2. In hysterical paraplegia incontinence of urine is never present (HAMILTON). There may be retention or temporary suppression of urine.

3. No amount of help can keep the patient from staggering or falling when she attempts to walk (REYNOLDS).

4. The foot in walking is simply dragged along, and not swung, as in organic hemiplegia (TODD).

5. In all sudden cerebral palsies, the nails of the affected extremities cease to grow. In hysterical palsies, of one limb or both, whether paraplegic or hemiplegic, the rate of nail growth is unaltered. (WEIR MITCHELL).

GENERAL PARALYSIS OF THE INSANE.

This curious disease, long unknown in the United States, has of recent years been frequently observed in the Northern and Eastern States, but so far, rarely or not at all in the South and West. It is a disease of advanced life, whose pathognomonic characteristics are constant troubles of motility, a progressive loss of mental power, and a constant belief on the part of the patient that he is perfectly well, and in the enjoyment of magnificent fortune and gigantic powers (*delires des grandeurs*).

The following are the progressive traits of the disease as generally observed:

PSYCHICAL SYMPTOMS.—1. General restlessness and unsteadiness of mind, with impairment of attention; alternating with apathy and drowsiness.

2. A change in disposition and temper, and a general loss of self-restraint; at first as regards trivial social observances, and then as regards general conduct.

3. Impairment of the reflective powers, so that there is no logical and systematic development of thought.

4. General exaltation of thought, with a profusion of remembered images and ideas, and numerous extravagant desires.

5. Failure of memory and forgetfulness; at first of words, and then of events.

6. Delirious conceptions, and the transformation of desires into be-

liefs, these being generally connected with personal greatness and power.

7. Hallucinations of the senses, in which remembered sense impressions are so vivid and intense as to spread to the periphery.

8. Maniacal restlessness and excitement, in which present impulses and feelings instantly pass over into action.

9. Increased mental weakness, with the incoherent and fragmentary repetition of the false ideas previously entertained.

10. Failure of the senses, with more marked impairment of memory.

11. Complete fatuity, passage into coma and death.

MOTOR SYMPTOMS.—1. Persistent contraction of the occipito-frontalis muscle, and some dilatation of pupils, causing the eyes to be widely opened and the forehead wrinkled, and giving an expression of surprised attention to the face.

2. Persistent contraction and frequent tremors of the zygomatic muscles, giving a pleased and benevolent expression of countenance.

3. Slight muscular restlessness and unsteadiness.

4. Impairment of the power of executing fine and detailed movements, so that manipulative skill is lost while movements *en masse* are still well performed.

5. Fibrillar tremors of the tongue, and some loss of control over its movements, so that it is protruded with difficulty; is rolled about when protruded, and is suddenly withdrawn.

6. Twitchings of the nostrils and upper lip, with frequent tremors of the latter.

7. Impairment of articulation, which is thick and wanting in distinctness.

8. An alteration in the voice, as well as thickness and hesitancy in speech.

9. Loss of control over the combined movements of the hand and wrist, so that the handwriting generally deteriorates.

10. Changes in the pupils, which are at first irregularly contracted, and then become irregularly dilated.

11. An alteration in gait, which becomes unsteady; the more complex movements of the thigh, leg, and foot, and the balancing of the pelvis on the hip joints, being performed with difficulty.

12. General muscular agitation and restlessness.

13. Gradual loss of power in the muscles of the face, tongue, neck, and limbs.

14. Spasmodic contraction of the masseter muscles, causing grind-ing of the teeth.

15. Convulsive seizures—most marked on one side of the body, and followed by transitory hemiplegia.

16. Loss of control over the sphincters.

17. Complete prostration of muscular strength and helplessness, and difficult deglutition.

18. Contractions of the muscles of the limbs, and paralysis of the muscles of respiration.

The main diagnostic difficulty is to distinguish this from some phases of locomotor ataxy. The differences are that in general par-alysis the mental symptoms are always present, and always precede the motor phenomena. The first symptoms in general paralysis are chiefly cerebral; viz., mental excitement, great garrulity, noisy hilar-ity, bragging, early violence of behavior, and very usually some ex-hibitions of libidinous conduct; on the subsidence of excitement, the mind is found to be weak, and the motor phenomena gradually make their appearance.

In ataxia, the commencement is in the spinal functions. There is first an attack of pain of some remote part, occurring most frequently in the lower extremities, and dating several years back, considered at the time perhaps to be rheumatic; this pain is worse toward evening, or when the patient is not mentally occupied; it may improve or dis-appear for a time and return. Then follows a slight degree of numb-ness of the part; the patient feels as if he had trod on wool; occa-sionally " pins and needles " attack the part; in fact, those phenomena which we have all experienced after sitting in an awkward position, when one's own leg has "gone to. sleep." There is, as most of us know, want of feeling, want of recognition of the member, especially as to its size, and even its ownership, then atrocious pain, and formica-tion or "pins and needles." In this disease, on the subsidence of the pain, the patients exhibit some awkwardness in gait; the ataxy or want of order in the movement is evident, while vigorous muscular

movements can still be executed. These symptoms may extend over
ten or twelve years with very little change, except, perhaps, increasing
awkwardness in gait; there is doubtless some numbness of the cutaneous surface in the course of the disease; the phenomena appear to
spread upward by involving the functions of the nerves higher up;
the power of erection of the penis, and soon after the sexual appetite,
are lost, and, as the disease ascends, the expulsory power of the bladder and rectum become impaired. All this occurs in most cases while
little change takes place in the mental functions; but in other cases
the mind appears imbecile, the memory is affected, and there is distinct
alteration in behavior and conduct; but there are no lofty ideas, no
excessive excitement and garrulity, and in no case paroxysms of violence, or libidinous ideas.

The differences may be better seen in a tabulated form:

GENERAL PARALYSIS.	LOCOMOTOR ATAXY.
Runs its course in a few years.	Is much slower usually, and may last ten or even twenty years.
Commences with mental symp-toms.	Commences with pain in distal nerves.
Is attended with libidinous ideas.	Is attended with abolition of sexual feeling.
The motor symptoms are sec-ondary in the order of time.	The motor symptoms are the primary phenomena.
Is only rarely complicated with pelvic difficulties.	Pelvic symptoms are a promi-nent feature.
There often is great violence of conduct.	The mental phenomena are im-becility and impaired memory.

There is also a form of general paralysis due to syphilis. The differential features of this variety have been clearly defined by Dr. E. C.
SEGUIN.* We do not obtain the regular gradations and stages of the
true disease. The moral perversion which is peculiar to general paralysis is absent, neither do we see the pure, exalted notions. The

* *Hospital Gazette*, September, 1878.

fibrillary tremors that are so well marked in general paralysis are not present here. The articulation is more mumbling in character. We, likewise, are apt to have a great deal of actual paralysis of cranial nerves or body in these cases. In true general paralysis, after attacks of hemiplegia, the patient regains his full strength, whereas this is not apt to occur in the syphilitic variety. The following table will perhaps show clearly the main differences:

TRUE GENERAL PARALYSIS.	SYPHILITIC GENERAL PARALYSIS.
Prodromic stage.	Absent.
Exalted notions, numerous and varied, and relatively exalted according to the position in life.	Rare or absent.
Speech is tremulous and jerky.	Speech is thick.
Tremor of hands and lips.	Absent as a rule.
Preservation of strength.	Paresis or actual paralysis.
Pupils are apt to be contracted.	Apt to be open or wide.
None.	Palsy of third or of other cranial nerves.
None.	Headache nocturnal.
Transient aphasic attacks.	More serious aphasic attacks.
Spontaneous remissions.	Progressive except under treatment.

Some other differences between the two conditions are as follows: The patient with syphilis has none of the cravings or abnormal appetites of the other; the latter feels an impulse to get drunk or to have an excess of coition. The tendency to excessive coition is absent in syphilitic paralysis, and, indeed, there is a marked loss of the virile power. The temperature changes are also absent in syphilis of the brain. The rise in temperature in general paralysis of the insane is very great, often reaching 103° in exacerbations. There is no rise of temperature in syphilis of the brain, except, perhaps, when the patient has hemiplegia from a large lesion.

The most important point is, that in syphilis there is a paralysis; in general paralysis there is irritation and incoördination without true paralysis.

SPINAL HYPERÆSTHESIA, SPINAL IRRITATION (SO-CALLED), AND SPINAL WEAKNESS.

This affection has been described by some writers as spinal hyperæmia, by others as spinal anæmia; again as spasms of the spinal muscles, and lastly as abnormality of the spinal cells. Some have denied its existence altogether; but in fact it is a distinctly defined and not unusual disorder. About five-sixths of the cases are neurotic females, and it is often associated with uterine or ovarian disease; and as often has some antecedent history of a blow upon or other slight injury to the spine.

Its symptoms are of the most varied kind, so much so that it may simulate almost every known ailment; but a *careful examination of the spine* will reveal its true character. The diagnostic rules laid down by Drs. WILLIAM and DAVID GRIFFIN, in 1834, who first described the disease, have never been improved upon. They are as follows.

1. The pain or disorder of any particular organ complained of is altogether out of proportion to the constitutional disturbance.

2. The complaints, whatever they may be, are usually relieved by the recumbent position, are always increased by lifting weights, bending, stooping, or twisting the spine; and among the poorer classes are often consequent to the labor of carrying heavy loads, drawing water, etc.

3. The existence of tenderness at that point of the spine which corresponds to the disordered organ, and the increase of pain in that organ by pressure on the corresponding region of the spine.

4. The disposition to a sudden transference of the diseased action from one organ or part to another, or the occurrence of hysterical symptoms in affections apparently acute.

5. The occurrence of continued fits of yawning or sneezing. These are not very common in the disease, but when they do occur, they may generally be considered as characteristic of nervous irritation.

To this it may be added that the tenderness may extend along the

spine generally, but is always greater in one or two spots. Gastric
symptoms, headache and languor are usually well marked in spinal
irritation; but there is neither muscular spasm, atrophy, paralysis (except hysterical) nor waist constriction, which serve to distinguish it
from a large class of spinal diseases.

In regard to the nature of *spinal irritation*, we know nothing definite
at present (ERB). WOOD declares that "there is not the slightest evidence of anæmia or congestion, or of any other recognizable spinal
irritation."* Lumbago, omodynia, scapulodynia, dorsodynia (forms
of myalgia), should be distinguished from spinal hyperæsthesia, though
often confounded with it (see page 136.)

NEURASTHENIA.

Neurasthenia is generally classed among the so-called "functional"
affections of the nervous system. BEARD, however, in his original
paper upon the subject, distinctly stated his conviction that there are
definite structural alterations, although perhaps undetectable, in the
nervous system, underlying the functional disorders which furnish the
varied manifestations of neurasthenia. That this neurasthenic condition may eventuate in various grave or even fatal disorders is now becoming more generally recognized, and the views expressed by Beard
nearly twenty years ago are daily obtaining more favorable consideration.

ARNDT† has proposed a new theory of the ultimate or underlying
cause of neurasthenia. According to his theory, neurasthenia is in
all cases due to a defective nervous organization, the nervous supply
being inadequate to perform the functions demanded by the physical
organism, of which it is respectively a part. In some instances the
disproportion between •nervous and non-nervous tissue may be so
slight that no disorder results until excessive demands are made upon
the nervous power, or until some grave interference with general nutrition—*i. e.*, some depressing disease, injury, or other cause of mal-
assimilation—makes the inequality more prominent. These are the
cases which usually manifest themselves decidedly for the first time in

* Nervous Diseases and their Diagnosis. Phila., 1887, p. 284.
† Die Neurasthenie, Wien, 1886.

advanced life in the broken-down business and professional man, bankrupt of nerve-force, of which this country has so large a proportion, and which seem to justify BEARD'S designation of neurasthenia as a distinctively American disease. So long as food is well digested, the blood properly oxygenated, the products of retrograde metamorphosis removed from the system by normally-acting emunctories, and the brain allowed intervals of recuperating rest, the individual does not become neurasthenic. But when any one or more of these conditions fail of attainment, the shortcomings of the defective nervous organization become apparent after a shorter or longer interval.

If, on the other hand, the original disproportion between the nervous system and the rest of the organism is great, the evidences of neurasthenia are early manifested—either in infancy, in childhood, at puberty, or during adolescence.

The trivial importance attached to neurasthenia by the majority of general practitioners often results in serious harm to the afflicted individuals. The diagnoses of hysteria, hypochondria, or syphilophobia, so frequently heard in our clinical amphitheatres, are responsible for much of the flippancy with which many of these heavily-burdened patients are treated. Lightly to accuse a woman of being " hysterical," or a man of being " hipped," may often be the gravest injustice, attaching, as it does, a stigma to the patient which may lead to permanent valetudinarianism, or even to self-destruction. Dr. EMMET has shown, by his brilliant clinical studies upon the consequences of neglect of certain lacerations of the cervix uteri, how many cases of so-called hysteria are dependent upon a purely physical condition of the nerves involved in the laceration. In the light of present knowledge, the nervous symptoms of laceration of the cervix uteri must be considered as expressions of the lack of nerve-power.

The manifestations of neurasthenia, diverse as they are in causation, are no less frequent than bewildering in their multiformity. Cases now and then occur characterized by general indisposition, vague nerve, muscle, or bone-pains, loss of appetite, sleeplessness, depression of spirits, frequent attacks of sick-headache, lack of the power of sustained intellectual application, and similar symptoms not referable to any special organ or apparatus. In other cases the languid or semi-

paretic condition is replaced by nervous exaltation or irascibility.   In
fact, both conditions may characterize the same case at different times.

These cases might with propriety be called cerebro-spinal neurasthenia.   They are exceedingly troublesome to manage, and in many
instances terminate in insanity or structural disease of the spinal cord.
The gradual change from soundness to unsoundness of cerebral and
spinal functions in these cases is well described by BLANDFORD* and
ARNDT.†

The most marked forms of cerebro-spinal neurasthenia result from
extreme nervous exhaustion from any cause.   Intense intellectual application in persons with a neurasthenic predisposition—in other
words, with a primarily defective nervous organization—frequently
determines an outbreak.   Many of the leading writers, artists, and
public men of the day belong to this class.   Among the more prominent neurasthenics of the present century were Walter Savage Landor,
Carlyle, Dante Gabriel Rossetti, Richard Wagner, and Victor Hugo.
In this country two of the most distinguished women writers have
recently suffered to an extreme degree from cerebro-spinal neurasthenia.

Many of these cases can be immediately traced to some traumatism,
an especially large number being the consequence of railroad injuries.
Mr. ERICHSEN and Dr. BEARD have both described a number of such
cases in detail.   In the sad case of the American poet, John G. Saxe,
now a confirmed victim to melancholia, the origin of the disease was
a spinal concussion received in a railway accident.   The methods of
studying the brain recently developed by M. Luys will doubtless enlarge our knowledge of the more intimate nature of the processes concerned in the evolution of morbid psychic and motor manifestations
resulting from traumatism.

Neurasthenia as effect, and probably in a measure also as cause,
of local alterations of nutrition, is extremely frequent.   The manifold
reflex symptoms of disturbances in the genito-urinary system are simply manifestations of neurasthenia.   Lallemand, Acton, Beard, S. W.
Gross, Moreau, Howe, Hammond, Von Krafft-Ebing, Tarnowsky,

* "Insanity and its Treatment," Lecture III.
† Loc. cit., pp. 171–202.

Ultzmann, and, above all, Otis, have shown the very intimate connection between genito-urinary diseases in the male and nervous exhaustion.   Not merely did the profound original researches of Otis point out the connection between urethral stricture and many nervous symptoms previously misunderstood, but his practical ingenuity gave to the profession a ready means of cure.   Dr. Sayre has shown the etiological relations of adherent prepuce to epileptoid conditions, and has likewise indicated the methods of relief.   In the writer's experience, an elongated and phimotic prepuce has not rarely seemed to be the cause of the train of symptoms comprised under the designation spermatorrhœa.   At all events, the simple dorsal division of the prepuce has seldom failed to check promptly the frequent involuntary emissions and render the nervous symptoms amenable to appropriate treatment.   In one marked case of pathophobia, under treatment two years ago, the symptoms seemed to be attributable to a frequently-recurring herpes-preputialis.   At all events, improvement in the psychic condition progressed *pari passu* with relief of the local disorder.

The female sex is probably subject to neurasthenia from sexual disorders to a still greater degree than the male.   Emmet's epoch-marking investigations into the consequence of laceration of the cervix uteri show most conclusively the powerful influence of local nerve-lesions is the production of neurasthenia.   Trustworthy authorities have stated that mental alienation may result from a continuance of marked neurasthenic symptoms.   During a term of service in the Maryland Woman's Hospital and in many cases in private practice, the writer has had opportunities to observe the rapid disappearance of the symptoms of nervous depression after appropriate local treatment.

Again, the manifold neurotic symptoms which attend attacks of chronic oöphoritis or pelvic peritonitis and cellulitis, and which not rarely eventuate in insanity, are evidences of neurasthenia depending upon local pathological conditions.   A recent clinical contribution by MORITZ MEYER,* calls attention to the frequency with which certain neuroses are dependent upon local neuritic affections, and points out the appropriate remedy.

---

* *Berliner Klin. Wochenscrift*, October 25, 1886.

9

Mr. H. B. HEWETSON* has also reported a number of cases of general neuroses due to eye-diseases. It is, of course, not pretended that all the cases of general neurotic disorder depending upon local affections are properly called neurasthenia; but the relation is much more frequent than is generally believed.

On the other hand, a neurotic or neurasthenic constitution may be responsible for the intensification, if not also for the origin, of the local disease. This is unquestionable in certain of the disorders of the genito-urinary apparatus in both sexes. Involuntary seminal emissions and functional impotence are notoriously frequent consequences of a lack of nerve-force. In women, ovarian irritability, amenorrhœa, and dysmenorrhœa can often be referred to the same source.

It looks very improbable at first sight that new growths, especially cancer, should be rationally attributable to the influence of nervous depression. However, so acute an observer as Sir JAMES PAGET seems to incline to the view that such a relation may possibly exist. SNOW† states that of one hundred and forty cases of mammary cancer analyzed by him, "one hundred and three gave an account of previous mental trouble, hard work, or other debilitating agency." Of one hundred and eighty-seven cases of uterine cancer, nearly one-half (ninety-one) gave a similar history. A number of detailed histories given by this author seem to bear out the claim that mental depression (neurasthenia) is a frequent cause of cancer. ARNDT‡ is positive that neurasthenia is an etiological forerunner of cancer, and relates three illustrative cases. WILLARD PARKER‖ states that "there are the strongest physiological reasons for believing that great mental depression, particularly grief, induces a predisposition to such a disease as cancer, or becomes an exciting cause under circumstances where the disease had already been acquired." In over one-fourth of PARKER'S cases of cancer of the breast (one hundred and six out of three hundred and ninety-seven), nervous depression or conditions directly leading to it are set down as forerunners of the disease.

* *London Lancet*, November 27, 1886.

† "Clinical Notes on Cancer," London, 1883, p. 30.

‡ *Die Neurasthenie*, Wien, 1886, pp. 202–208.

‖ "On Cancer," New York, 1885, pp. 40–43 and 58.

Cooke* points out the fact that in women cancer occurs on an average ten years earlier than in men, coinciding in the former sex with the menopause, and in the latter with the period of greatest exhaustion of the nervous system. He remarks that "the women of fifty are very generally more worn and wasted than the men of seventy." It is between the fiftieth and sixtieth years that neurasthenia from professional overwork is most frequent. The merchant, official, or professional man is reminded that he cannot do the same amount of work at fifty that he could at thirty ; in the words of a distinguished Philadelphia teacher of the last generation, " he cannot get a three-minute pace out of a four-minute horse."

Jonathan Hutchinson † says upon the same point : " We shall probably be not far from the truth if we admit senility of tissues, local or general, to be the one predisposing cause of cancer with which we are acquainted. . . . . A little step further may next be taken, in the belief that everything which hastens senility, either local or general, will increase the predisposing influence, and *in this category may be placed anxiety, distress, overwork, and excesses of all kinds. It is highly probable that under such conditions a state involving increased proneness to cancer may be induced,* and, further, *that if offspring are produced after that state has been developed, they will inherit that tendency."*

Parker, in commenting upon the unfriendly attitude of modern microscopical pathologists towards the theory that derangements of the nervous system may determine the origin of cancer, makes the wise observation that too close attention to the revelations of the microscope may obscure our view into the conditions underlying the visible changes. The question certainly deserves more unprejudiced study than it has heretofore received.

Disorders of the digestive system are intimately related to neurasthenic conditions. Even in persons with no demonstrable inherited nerve-weakness, any great depression of the assimilative organs results in a passing neurasthenia. Gastric or intestinal catarrhs, or disorders of the liver, or an overloaded condition of the large intestine,‡

* " On Cancer," London, 1865.

† " The Pedigree of Disease," New York, 1885, p. 74.

‡ J. S. Jewell, in *Neurological Review*, vol. i., No. 4, p. 218 *et seq.*

if long continued, are nearly always accompanied by neurasthenia. Where there is an inherited neurasthenic constitution, the digestive troubles are doubtless sometimes the consequence of a foregoing nervous depression. Every practitioner sees such cases almost daily; but the relations between cause and effect often remain unrecognized. The sub-diaphragmatic self-consciousness of Carlyle was doubtless of this nature. The loss of digestive power, insomnia, mental irascibility, and unmanly treatment of his wife and friends, which characterized his later life, all point to an inherited neurasthenic predisposition. Some of these characteristics were still more pronounced in Walter Savage Landor.*

The frequency of the diagnosis of " nervous dyspepsia" is an indication that a tendency to attribute to the nervous system a share in the production of painful digestion is gaining ground. However, as Dr. Clifford Allbutt clearly pointed out in the Gulstonian Lectures for 1884,† what is so commonly called nervous dyspepsia, is not really dyspepsia or gastric disorder at all, but a manifestation of neurasthenia, curable only by appropriate treatment directed to the nervous system. Drs. S. Weir Mitchell ‡ and Richard M. Hodges ‖ have also pointedly called attention to the customary lack of accuracy in the diagnosis of these cases, and have given many judicious hints for their management.§

In making a diagnosis, these characters will be distinguished from those of disease of the vertebræ by careful physical examination, and the age, sex, and history of the case.

HYSTERIA.

Few diseases present at times greater difficulties of diagnosis than this protean complaint. Its counterfeits of various maladies will be considered elsewhere (see the Index); at present we shall seek for a pathognomonic symptom of the general condition.

* " Landor," by Sidney Colvin. New York, 1881.
† " The Visceral Neuroses," Philadelphia, 1884.
‡ " Fat and Blood," and " Nervous Diseases of Women."
‖ *Boston Medical and Surgical Journal,* July 10, 1884.
§ George H. Rohé, *Philadelphia Med. Times,* vol. xvii., p. 284.

One is offered by Dr. THOMAS BARLOW.* Rejecting as unsatisfactory all statements depending upon the patient's veracity, he finds a diagnostic test in the presence of *analgesia*. If, *while the patient's attention is directed to something else*, a needle be introduced into the forearm, and no wincing occurs, there is the strongest presumption that we have to do with a case of hysteria. Again, it has been long known that hysterical patients are often extremely tolerant of laryngoscopic examination. Great advantage will be found in examining a presumed hysterical patient's larynx, and thus fixing her attention while somebody at the same time inserts a needle in her forearm. Absolute tolerance of these two simple methods of examination is quite decisive.

Another characteristic relates to the pain so frequently complained of. While it is stated to be exceedingly acute, and the part tender to the slightest pressure, if the attention of the patient is engaged, very firm pressure may be made without the patient wincing. Moreover, there is noted very often a co-existence of severe pain in the epigastrium, the left side and spinal column—the *trepied hystérique*, or hysterical tripod of French authors.

The *globus hystericus*, a sensation of a foreign body in the throat caused by spasmodic contraction of its muscles, is a common symptom. The urine may be suppressed, or may be limpid and watery, and of unusual quantity.

If with these traits are united youth and female sex, and fairly maintained nutrition; ovarian or uterine disturbance; the general symptoms harmonious and exaggerated; the mind clear; and the disappearance of contractions, etc., under anæsthesia; the diagnosis is complete.

The most serious mistake would be the confounding of a hysterical paroxysm with an epileptic fit. The following table of distinctions between the two is given after CHARCOT and DA COSTA:

EPILEPSY.	HYSTERIA OR HYSTERO-EPILEPSY.
Sudden and complete loss of consciousness. Patient liable to injure herself in falling.	Gradual or only partial or apparent unconsciousnes. Patient rarely falls so as to injure herself.

* *Medical Times and Gazette*, February, 1878.

EPILEPSY.

Livid face; escape of frothy saliva from the mouth; eyelids half open; eyeballs rolling; grinding of the teeth; biting of the tongue; more or less insensibility of the pupils to light.

Distortion of countenance.

Patient evinces no feeling.

Aura epileptica of short duration.

Convulsions often more marked on one side than on the other; and more tonic than clonic. Agitation maniacal and disorderly.

Paroxysms generally of short duration.

Paroxysm followed by a heavy, half comatose sleep, by headache, and dullness of intellect. Stertor. No hallucinations.

Frequently occurs at night.

No particular connection with uterine disturbance: although a paroxysm often takes place at the menstrual period.

HYSTERIA OR HYSTERO-EPILEPSY.

Face flushed or complexion unaltered; no froth on lips; eyelids closed; eyeballs· fixed; neither grinding of the teeth nor biting of the tongue; pupils react readily.

No distortion of countenance.

Patient sighs, or laughs, or sobs.

Aura often prolonged one or two days. Globus hystericus.

No such differences; convulsions clonic. Agitation emotional, often *en pose.*

Paroxysms generally of longer duration.

Paroxysm not followed specially by sleep; patient often, after attack, wakeful and depressed in spirits. Little or no stertor. Hallucinations.

Rarely occurs at night.

Often connected with disorders of the uterus, or of menstruation.

The distinctive characters of tetanus, convulsive hysteria, and similar diseases are shown in the following table:*

TETANUS.	HYSTERIA.	POISONING BY STRYCHININE.	TETANY.
Commences with malaise, muscular twitchings, rigidity, and tenderness.	Blindness and weakness may usher in an attack.	Exhilaration and restlessness.	Numbness and tingling in fingers and toes, followed by tonic spasm of limb.

* Partly from H. C. Wood's Therapeutics. Sixth Edition, Phila., 1886.

TETANUS.	HYSTERIA.	POISONING BY STRYCHNINE.	TETANY.
Muscular symptoms gradually become more prominent. Lock-jaw an early symptom.	Not limited to certain muscles; jaw not permanently set.	Early muscular twitchings or general convulsions. Jaw affected late.	Muscular spasm in one limb, or several successively, of brief duration usually, may last an hour or two; rarely extends to muscles of jaw.
Persistent rigidity with tendency to opisthotonos, empos thonos, pleurosthotonos, or orthotonus.	Opisthotonos persistent, and intense rigidity between convulsions.	Muscles relaxed between paroxysms; soreness and stiffness after convulsions have ceased.	Relaxed. Paroxysms recur at variable intervals, may last for weeks or months.
Consciousness unaffected.	Lost during convulsions.	Generally preserved during convulsions, unless asphyxia should occur.	Consciousness unaffected.
Draughts of air, sudden noises, etc., excite convulsions.	Not the case.	The slightest breath of air causes convulsion.	Spasms brought on by pressure upon nerve-trunks or blood-vessels.
Complains of pain.	Hysterical sobbing and crying.	Screams with pain when convulsions occur.	Less painful.
Eyes open and staring through convulsions.	Eyes closed.	Eyes widely opened.	Facial muscles rarely involved.
Recovery rare, generally fatal in first attack.	Recovery the rule; attacks likely to recur.	Attack single, and likely to be fatal.	Prognosis usually favorable, the attacks subsiding after a few months. Occasionally death occurs from asphyxia.

NEURALGIA.

The positive signs which distinguish a case of neuralgia, are succinctly set forth by Dr. FRANCIS E. ANSTIE,* as follows:

(1) The first and most essential characteristic of a true neuralgia is that the pain is invariably either frankly intermittent, or at least fluctuates greatly in severity, without any sufficient and recognizable cause for these changes.

(2) The severity of this pain is altogether out of proportion to the general constitutional disturbance.

(3) True neuralgic pain is limited with more or less distinctness to a branch or branches of particular nerves; in the immense majority of

* "Neuralgia and its Counterfeits," p. 565.

cases it is unilateral, but when bilateral it is nearly always symmetrical as to the main nerve affected, though a larger number of peripheral branches may be more painful on one side than on the other.

(4) The pains are invariably aggravated by fatigue or other depressing physical or psychical agencies.

These four cardinal points of the diagnosis may be further supported by the history of the patient. Either (1) he has previously been neuralgic, or liable to other neuroses, or comes of a neurotic family; or (2) there has been malarial poisoning of the blood; or (3) there has been some long-continued peripheral or central irritation; or finally (4), there has been constitutional syphilis.

The pains with which neuralgia is most likely to be confounded, are those arising from myalgia, spinal irritation, locomotor ataxia, cerebral abscess, alcoholism, syphilis, chronic rheumatism, and latent gout.

In comparing the pains of myalgia and neuralgia, the following are the more important points:

NEURALGIA.	MYALGIA.
Follows the distribution of a recognizable nerve or nerves.	Attacks a limited patch or patches that can be identified with the tendon or aponeurosis of a muscle which, on injury, will be found to have been hard worked.
Accompanies an inherited or acquired nervous temperament which is obvious.	Often occurs in persons with no special neurotic tendency.
Is usually not much or at all aggravated by movement.	Is inevitably and very severely aggravated by every movement of the part.
Is at first not accompanied by local tenderness.	Distinguished from the first by localized tenderness on pressure as well as on movement.
Painful points, when established in a later stage, correspond to the emergence of nerves.	Tender points correspond to tendinous insertions and origins of muscles.
Pain not materially relieved by any change of posture.	Pain usually completely and always considerably relieved by full extension of the painful muscle or muscles.

Treatment also offers a diagnostic means. The pains of myalgia will sometimes disappear at once by retaining the affected muscle at full extension, surrounding it with moist warmth, and giving 20 or 30 grains of muriate of ammonia internally.

Spinal irritation (spinal hyperæsthesia) is almost exclusively confined to women. There are nearly always hysterical symptoms, marked superficial tenderness over large portions of the surface, often merely cutaneous and becoming less acute with firm pressure. There are nearly always tender spots along the spine, and also over the epigastrium and the left hypochondrium.

Locomotor ataxia is mentioned elsewhere, and its symptoms described in sufficient detail. (See page 108.)

Alveolai abscess is often in its incipient stage mistaken for neuralgia.

Cerebral abscess, though rare, may give rise to a regrettable mistake, especially in its early stages, where severe pain in the head is almost the only conspicuous symptom. At this period the diagnosis from neuralgia must rest on the following points of contrast:

CEREBRAL ABSCESS.	NEURALGIA OF THE HEAD.
Often occurs secondarily to caries of internal ear, and purulent discharges, the result of scarlet fever, measles, etc., in childhood.	Rarely appears before puberty. No local assignable cause.
Frequently follows a blow or injury.	Comparatively seldom caused by a blow or other external injury, or caries of the bone.
No true "points douloureux."	If severe, soon presents, in most cases, the " points douloureux."
Usually the pain does not completely intermit.	Intermissions of pain complete and of considerable length.
Pain often excruciating from a very early period.	Pain usually not very violent at first.
Pain often limited in situation, seems deep-seated, though as often as not it has no relation to the site of the abscess.	Pain superficial; follows distribution of recognizable nerve branches belonging to the trigeminus or great occipital.

CEREBRAL ABSCESS.	NEURALGIA OF THE HEAD.
No well localized vaso-motor or secretory complications.	Usually lachrymation, or congestion of the conjunctiva, etc.
Very rare in old age; then usually traumatic.	Severe neuralgia is commonest in advanced life.
Relief from stimulant narcotics very transitory.	Relief from opium, etc., is much more considerable and permanent.

The pains of *chronic alcoholism* often closely stimulate those of true neuralgia. The habits and history of the patient, when known, point to the true origin of the suffering; also the insomnia, loss of appetite, foul breath, furred tongue and haggard countenance of the drunkard; and especially that the pains complained of *encircle the limbs near the joints*, rather than run longitudinally the course of the nerves in the limb, are all significant.

The osteocopic pains of *syphilis* are usually symmetrical; are aggravated by the warmth of the bed; are generally referred to the superficial bones, and do not exist long without some other and decisive symptoms of the poison manifesting themselves.

Chronic rheumatism and gout are each attended with such marked collateral disturbances that the suspicion of their presence can readily be set at rest or sustained.

MULTIPLE NEURITIS.

Dr. W. R. GOWERS, in his recent work (" A Manual of Diseases of the Nervous System "), gives a very clear description of this disease, which is now attracting so much attention.

He thinks that the " discovery " of multiple neusitis is one of the most important steps in the recent advance of neurology. The morbid change may be interstitial (between the fibres), or parenchymatous (degenerative change of the fibre itself), or both. He differentiates five forms of the disease.

(1) Diphtheritic neuritis: a parenchymatous degeneration of the nerves, associated, however, with disease of other parts. ·

(2) Tabetic neuritis: the slow chronic changes in nerve-trunks which occur in locomotor ataxia, and which may possibly not be inflammatory.

(3) Leprous neuritis: the slow interstitial overgrowth of connective tissue which occurs in anæsthetic leprosy.

(4) Endemic neuritis, as in the Japanese disease called kakké (and the more familiar beri-beri?).

(5) Under this head he places the "other forms of multiple neuritis," which, indeed, constitute the disease to which the name is usually restricted. Gowers finds the disease most common in females, which is unexpected when he states that by far the most common cause of it is chronic alcoholism. In fact, he uses the terms "alcoholic paralysis" and "multiple neuritis" as almost identical." Among other causes, however, which he admits to be sometimes operative are cold, fatigue, rheumatism, phthisis, septicæmia, and exhausting diseases.

The pathology may consist either in parenchymatous or interstitial changes in the nerve-trunks, but in severe cases probably in both. He insists that the fibres are always damaged; but in some cases the connective tissue presents little change. As the trunks are traced upward, the evidence of inflammatory changes becomes slighter, and in all forms the anterior roots are usually normal. The nerves in the limbs alone suffer in the majority of cases, and the affection is bilateral. The nerves supplying the extensor muscles of the hand and foot are especially liable to attack. Very rarely the phrenic, pneumo-gastric, facial, and hypoglossal have been inflamed, the other cranial nerves escaping. The muscles usually degenerate, as in simple forms of neuritis. The cord is usually normal. The symptoms are the same as follow inflammation of a single nerve, differing only in their wide range; they are especially motor palsy and sensory irritation. The motor palsy, as said above, is apt to be most pronounced in the musculo-spiral and anterior tibial nerves. The symptoms are apt to set in acutely, and to be accompanied by rigors and fever. The sensory symptoms come first as vague "pins and needles sensations" in fingers and toes, preceded perhaps by rheumatoid pains. These pains increase until finally the nerve-trunks become tender on pressure, and superficial nerves, as the ulnar may be felt to be distinctly swollen. The muscles also become very sensitive to pressure. Motor palsy supervenes and resembles in the arms lead palsy, with the important distinction that the supinator

longus muscle is apt to be involved, as it is not in plumbism. In
severe cases the flexors of the wrist and ankle (hand and foot) also
become involved, as do also the muscles moving the elbow-, knee-,
shoulder- and hip-joints. Electro-diagnosis is important. The nerve-
trunks lose their electrical excitability [probably more completely
than in some forms of anterior poliomyelitis], and the muscles put on
the reactions of degeneration. Gowers calls attention to a distinct inco-
ordination of movement, which it is important to distinguish from
locomotor ataxia. The sensory phenomena are, briefly, hyperæsthesia
of the extremities, especially of the palms and soles ; anæsthesia along
the radial side of the forearms and outer side of the leg in particular,
sometimes associated with pain (*anæsthesia dolorosa ?*). Reflex action
is *always* lost. Trophic changes and œdema have been observed.
The duration of the disease varies, but at best it is apt to extend over
several months. In very severe cases death has occurred from failure
of the muscles of respiration, especially the diaphragm. An important
complication is mental impairment, often with delusions, particularly
in the alcoholic cases.

INSANITY.

The principal forms of insanity are commonly considered under the
head of Mania, Monomania, Melancholia, Dementia, Idiocy and Imbecility ; to which may be added emotional and impulsive insanity, which
is now usually admitted. There is no ground for recognizing as a
distinct variety, moral insanity (FLINT), but moral degradation and im-
moral conduct occur in different forms of insanity, and may constitute
the earliest observed symptom, as in general paralysis.

The pathological anatomy of insanity has been thus stated by
SEGUIN :

I. Acute Recent Cases.	Vascular changes . . . . { Anæmia. / Hyperæmia. / Serous effusion. Old congestion. Hemorrhages into perivascular sheaths. Changes in gray matter (not demonstrable under the micro-scope.
II. Chronic Insanity and True Dementia.	Congestion of anæmia. Atheroma of vessels. Membranes changed, diseased and thickened. Nerve elements degenerated and atrophied. General atrophy of convolutions ; most marked in anterior con-volutions.

III. General Paralysis. { Similar to preceding, but differs in distribution of lesions; particularly in neuralgia; also granular ventricular changes. Often lesions in spinal cord co-exist. Subarachnoid hemorrhage not a frequent charge.

IV. Syphilitic Insanity.

In mania, chronic alcoholism, and general paralysis, *pachymeningitis hemorrhagica* is often found. Section through the thickened dura reveals alternate layers of tissue and coagulated blood.

Patients under twenty-five years of age seldom have chronic insanity; when they do, the vascular changes are less marked; in older patients the vessels become fatty and atheromatous. The capillaries show fatty change, their nuclei being first affected; in the arteries the muscular coat becomes fatty. Through large tracts of brain, granular and amyloid bodies are found. Old inflammatory changes in the meninges are quite common. The arachnoid is opaque in spots.

The causes of insanity are very complex; the PHYSICAL CAUSES are thus given by the same author:

Injuries to head . . . . {
Concussion.
Hemorrhages.
Meningitis.
Depressed fracture.
Abscess.
Tumor.

Syphilis. . . . . {
Gummy formations.
Arteritis.
Meningitis, etc.

Dyscrasiæ. . . . . . . {
Malaria.
Narcotic poisoning, etc.

Alcohol. . . . . . . {
Delirium ebriosum.
Delirium tremens.

Sexual excess . . . . . {
Individual venery.
Masturbation.

Peripheral irritation. . . {
Uterine difficulty.
Vaginismus.
Masturbation.

The MORAL CAUSES are misery, depression, emotions, excitement, remorse, fear, grief, religious fervor, excess of joy, the spirit of speculation, etc.

The two forms, *mania* and *melancholia*, have their general distinctions, as follows:

MANIA.	MELANCHOLIA.
Ego elated and over active.	*Ego* is depressed and does not react normally on external world.
Joy and excitement generally previal, sometimes comic emotions characterize attacks.	Sadness and fear; religious feelings strongly developed.
	Reduced ideation. { Few motions, and even absolute silence
Over-ideation and over-action. Resulting therefrom incoherence and delirium and violent acts; general restlessness.	Reduced action . { Immobility relative or total, and even cataleptoid state.
Insomnia.	Insomnia (less marked).

PHYSICAL SYMPTOMS.

Increased circulation.	Lessened circulation.
Increased calorification.	Lessened calorification.
Increased (?) assimilation.	Lessened assimilation.
Increased voracity.	

The earliest symptoms of insanity are a marked change in the habits of the individual; proneness to irritability and loss of self-control; an alteration in the emotions; failure of memory; untidiness of dress; insomnia and disturbing dreams; unusual loquacity or taciturnity; defective reasoning; accepting as real various fancies and illusions; a furtive watchful air; groundless suspicion of those around, especially of infidelity to marital vows on the part of wife or husband. In combination with these mental symptoms, the pupils are often dilated, frequently irregular and sluggish in obeying the stimulus of light; and a pulse hard, rapid, and variable, 100 or over, a pulse which is not equal in both wrists (HENRY HOWARD). The tongue is pasty, the breath foul, and the bowels constipated. The digestion is impaired, and the appetite irregular and capricious. There is encroachment of the senses upon each other; the sense of sight, for instance, is substituted by audition, and the patient will describe scenes with the greatest minuteness of detail as occurring in the neighborhood, even to the color of clothing, and peculiarities of appearance which could only be learned by inspection, but which he knows, because he "heard the dogs barking," or the noise of some fancied tumult. The hearing of inaudible voices calling opprobrious names is even popularly recognized as indicating insanity. Where no organic changes can subsequently be detected in the brain, we are forced to the conclusion that there is either some non-recognizable defect, or that there must be functional disorder of the brain, as of other organs in the economy.

H. C. Wood * divides complete non-periodic insanities into three groups on the bases of the emotional conditions, thus affording an easy means of recognizing clinical symptomatic groups representing affections of whose pathology we have no distinct knowledge :

EMOTIONAL STATE.	FORM OF INSANITY.			
Exaltation.	Mania. { Acute. Chronic.			
Depression.	Melancholia. { Melancholia. Katatonia.			
Apathy from loss of activity, normal or variable.	Mental Deteriorations.	{ Imbecility. Primary Dementia. Terminal Dementia.	{ Organic. Developmental. Miscellaneous.	{ Senile. Hebephrenia. Shock. Primary. Confusional Insanity. Stuporous Insanity.

THE CLASSIFICATION OF MENTAL DISEASES.

On the 8th of September, 1886, a conference representing various distinguished scientific bodies of the United States and Canada, and eminent alienists and publicists, was held at Saratoga, New York, for the purpose of co-operating with an International Committee appointed at Antwerp last year, under the auspices of the Belgian Society of

* Nervous Diseases, page 471.

Mental Medicine. The object was to recommend to this International Committee a basis for classification of mental diseases, in the hope of securing a uniform basis in all countries of the civilized world. The following plan of classification was adopted:

1. Mania: acute, chronic, recurrent, puerperal.
2. Melancholia: acute, chronic, recurrent, puerperal.
3. Primary delusional insanity (monomania).
4. Dementia: primary, secondary, senile, organic (tumors, hemorrhages, etc.).
5. General paralysis of the insane.
6. Epilepsy.
7. Toxic insanity (alcoholism, morphinism, et.).
8. Congenital mental deficiency (idiocy, imbecility, cretinism).

This classification is remarkably free from the elaborate differentiations which are now adopted by some writers. It looks like a reversion to the simplicity of a former and, it was supposed, more ignorant generation. The truth is that while the clinical study of insanity has advanced and been very prolific of a useful and instructive literature, our knowledge of the *essential* functions and pathology of the cerebral masses is not greatly more advanced than it was with the fathers; and without more of such intimate knowledge (unbiassed by speculation and opinion) the very foundation for a perfectly acceptable classification does not exist.

CHAPTER II.
DISEASES OF THE RESPIRATORY APPARATUS.

DISEASES OF THE LARYNX.—*Symptoms of Laryngeal Diseases—Diagnostic Table of Acute Laryngitis, Chronic Laryngitis, Syphilitic Laryngitis, Tubercular Laryngitis, Perichondritis, Benign Growths, Malignant Growths, and Neuroses of the Larynx—Croup and Diphtheria: Spasmodic Croup, Inflammatory Croup, Membranous Croup, and Diphtheria—Tonsillitis, Catarrhal and Parenchymatous.*

DISEASES OF THE LUNGS.—*The Regions of the Chest—Normal Differences between the two sides of the Chest—Methods of Physical Examination—Normal Respiratory Sounds—Normal Voice Sounds—Abnormal Percussion Sounds—Abnormal Respiratory Sounds—Abnormal Voice Sounds—General Rules for Diagnosis—The Forms of Phthisis (Catarrhal, Fibroid, Tubercular)—The Diagnosis of Incipient Phthisis—Diagnosis between Incipient Phthisis and Bronchitis—Clinical History of Phthisis—Acute Phthisis—Syphilitic Phthisis—Bronchitis, Acute and Chronic—Capillary Bronchitis compared with Pneumonia—Pneumonia and Pleurisy—Pleurisy with Effusion and Pneumonia with Consolidation compared—Diagnosis between Pneumonia and Pulmonary Apoplexy—Pulmonary Embolism—Asthma—Pneumothorax and Pneumo-hydrothorax—Emphysema, Vesicular and Interlobular—Cancer of the Lung.*

In studying diseases of the respiratory apparatus, we find in addition to the disorders of the lung proper and its serous investment, the pleura, that there are associated organs which likewise may be the seat of disease; these are the bronchi and trachea, the larynx, pharynx, and upper air passages. We commence with:

DISEASES OF THE LARYNX.

The general symptoms of laryngeal diseases, together with their causes and examples, may be arranged in the following tabular form:

146

DIFFERENTIAL DIAGNOSIS.

A.—Functional, or Subjective.

SYMPTOMS.	ACUTE LARYNGITIS.	CHRONIC LARYNGITIS.
VOICE.	Hoarse, becoming aphonic.	Hoarse, uncertain, easily fatigued.
RESPIRATION.	Not embarrassed prior to œdema; then stridor, dyspnœa, and even apnœa.	Seldom embarrassed.
COUGH.	Dry, hard, shrill, metallic, aphonic; on exudation, moist.	Irritative, with slight expectoration of glutinous pellets.
DEGLUTITION.	Painful when œdema has taken place, or from associated pharyngeal inflammation.	Rarely affected.
PAIN AND ALTERED SENSATION.	Sensation of tightness and constriction; tender to external pressure.	Painless; sense of fatigue after vocal exercise.

B.—Physical, or Objective.

SYMPTOMS.	ACUTE LARYNGITIS.	CHRONIC LARYNGITIS.
COLOR.	Intense, uniformly increasing superficial hyperæmia; translucent on event of œdema.	Partial and modified submucous hyperæmia.
FORM AND TEXTURE.	Thickening and stenosis from œdema, loss of tissue rare, except in phlegmonous form.	Occasionally slight erosion, never ulceration, thickening or narrowing.
POSITION.	Unaltered.	Unaltered.

C.—Miscellaneous.

SYMPTOMS.	ACUTE LARYNGITIS.	CHRONIC LARYNGITIS.
EXTERNAL.	Pharynx usually synchronously implicated.	Pharynx usually synchronously implicated.

SYPHILITIC LARYNGITIS.	TUBERCULAR LARYNGITIS.
Secondary. Hoarse. *Tertiary.* Characteristicly raucous; seldom aphonic.	Sometimes aphonic in earlier stages; completely lost in advanced disease.
Secondary. Unchanged. *Tertiary.* Increasing embarrassment according to stenosis.	Early hurried; greatly embarrassed with advance of disease.
Secondary. Slight hacking. *Tertiary.* Infrequent, with but slight expectoration, unless peri-chondritis supervene.	Greatly influenced by amonnt of lung disease; painful. Expectoration variable; generally frothy.
Secondary. Varies with deposit on epiglottis or arytenoids. *Tertiary.* Often difficult; very rarely painful.	Extremely difficult and painful, from early period to termination.
Characteristic absence of pain except when cartilages are attacked.	Pain only experienced in functional acts
Secondary. Mottled, more or less symmetrical hyperæmia. *Tertiary.* Hyperæmia of portion attacked prior to ulceration; with infiltrated appearance.	Anæmia followed by opaque grayish color; margins of ulcers hyperæmic.
Secondary. Occasional superficial ulceration at vocal process; slight general submucous infiltration. *Tertiary.* Deep, circumscribed destructive ulcers, of yellowish color, followed by cicatricial narrowing, occasionally paralysis and new formations.	Solid submucous thickening of epiglottis and aryepiglottic folds, elevation and ulceration of racemose glands giving worm-eaten ulcers, which commingle and attack deeper tissues.
Secondary. Unaltered. *Tertiary.* Deformity from intrinsic cicatrices and pharyngeal outgrowths.	No displacement; tendency for thickcned parts to transgress boundaries of pharynx.
Secondary. Pharynx and skin genèrally recently implicated. *Tertiary.* Seldom synchronous implication, but usually scars of previous similar pharyngeal ulceration, possibly adhesion.	Lungs either primarily, synchronously, or subsequently involved. Generally anæmia, rarely ulceration of pharynx. General emaciation.

A.—Functional, or Subjective.

SYMPTOMS.	PERICHONDRITIS.	BENIGN GROWTHS.
VOICE.	Painful, easily fatigued, but not necessarily impaired.	Very variable, from slight hoarseness to complete aphonia, even in the same case.
RESPIRATION.	Variable according to cartilage attacked.	Seriously embarrassed in one-sixth of cases; depends on situation.
COUGH.	Generally early spasmodic; with caries characteristic. Purulent expectoration, unless abscess is encysted.	Generally limited to effort to dislodge foreign body; may be expectoration of atoms of growth.
DEGLUTITION.	Varying from dysphagia to aphagia, according to pressure on gullet.	Only impaired in rare cases, in which epiglottis or aryepiglottic fold is involved.
PAIN AND ALTERED SENSATION.	Pain variable with cause; most severe in gouty form, but not then constant.	Characteristically absent.

B.—Physical, or Objective.

COLOR.	Hyperæmia generally limited to portion attacked, sometimes extending to contiguous vocal cord.	Variable with nature of neoplasm; slightly increased vascularity of mucosa generally.
FORM AND TEXTURE.	Ulceration often absent, substituted by encysted abscess, causing narrowing, compression and paralysis.	Varies with situation, size and nature of growth, never ulceration. May cause narrowing and paralysis.
POSITION.	May be considerable alteration of supra- and infra-glottic space.	Position of normal parts seldom changed.

C.—Miscellaneous.

EXTERNAL.	Occasional constitutional manifestations.	Nil.

MALIGNANT GROWTHS.	NEUROSES.
Impaired by mechanical causes when invaded from pharynx; may be early lost in primary disease.	Lost in bilateral paralysis of adductors; impaired in other paralyses; not necessarily in spasm.
Early quickened on exertion; later paroxysmal dyspnœa from stenosis or compression.	Only embarrassed in paralyses of adductors and in spasmodic affections.
Not necessarily present; expectoration scanty; occasionally blood and portions of neoplasm.	Paroxysmal, when recurrent, is implicated and in spasmodic affections.
Always difficult and painful; often the earliest symptom.	But slightly impaired or unaffected.
Ever present and severe, extending upward to the ears, and to sympathetic glandular enlargements.	Only experienced when sensory system affected. Diminished sensation in motor paralyses and in anæsthesia.
Increasing localized vascularity tending to lividity in any part except vocal cords or ventricles, when neoplasm is whitish-gray or pale rose.	In paralysis of abductors, occasional vascularity of affected vocal cords.
May cause compression, narrowing and paralysis before ulceration, which is always accompanied by thickening. Extensive indolent gray, greenish, or almost black ulcers.	Form of glottis varying with nature of paralysis, without extrinsic thickening.
Early displacement, especially when invading from pharynx, and when neighboring glands enlarged.	Paralyzed cord not displaced, but often fixed in one position.
Glandular infiltration, but complete immunity of other organs of body from similar disease both prior and subsequent to appearance in laryngo-pharynx. General emaciation.	Sympathetic functional disturbances in other organs, or organic disease of cardiac or lymphatic system, or associated cerebral disease or chronic toxæmia.

SYMPTOMS OF LARYNGEAL DISEASES.

SYMPTOMS.	CAUSES.	EXAMPLES OF DISEASE.
Dysphonia.	Alteration in the vocal cords from thickening, ulceration, diminished tension, morbid growths, etc.	Acute and chronic laryngitis. Laryngeal phthisis. Papillomata, etc.
Aphonia.	Non-approximation of the vocal cords, either mechanical or due to paralysis of some of the muscles attached to them.	Cicatrization. Swelling of arytenoid cartilages. Tumors. Pressure on recurrent laryngeal nerves, etc.
Dyspnœa.	Narrowing of the orifice of the glottis.	Paralysis of muscles opening glottis. Laryngismus stridulus. Œdema, growths and cicatrices contracting rima glottidis, and pressure external to larynx.
Stridor.	Always accompanied by dyspnœa, and produced by the same causes.	As in dyspnœa.
Cough.	Irritation of the laryngeal mucous membrane, or the nerves of the larynx.	In most laryngeal diseases it is of a peculiar shrill, brazen character.

Laryngitis has been divided by some writers into the following forms:

Œdematous laryngitis. Diphtheritic laryngitis.
Catarrhal laryngitis. Syphilitic laryngitis.
Erysipelatous laryngitis. Tubercular laryngitis.
Croupous laryngitis. Exanthematous laryngitis.

TRAUMATIC LARYNGITIS.

Among authors who have paid especial attention to this subject, there are few that stand higher than Mr. LENNOX BROWNE, of Lon-

don, who in his work on the *Diseases of the Larynx* gives the diagnostic table presented in the preceding pages.

The chronic laryngitis of syphilis cannot with certainty be distinguished from the other forms of chronic laryngitis without inquiry into the history of the case; although a probable diagnosis may be made
where treatment by anti-syphilitic remedies is successful.

In tertiary syphilis there is deep and extensive ulceration, not necessarily preceded by thickening; the epiglottis is attacked early, the ulceration is often followed by cicatrization and contraction, causing
stenosis of the larynx.

In the study of laryngeal diseases the use of the laryngoscope is
indispensable to correctness of diagnosis. We take it for granted
that the practitioner is conversant with this instrument, and the proper
methods of employing it. It reveals the physical or objective local
symptoms, which are of much more value than the subjective ones
derived from the patient's statements.

CROUP AND DIPHTHERIA.

The general sign common to this class of diseases is a *laryngeal
stridor;* they are divided into those where there is a formation of false
membrane and where there is not.

Without false membrane.

Spasmodic croup or laryngismus stridulus.

Inflammatory croup, simple catarrhal laryngitis.

With false membrane.

True croup or membranous croup.

Diphtheria.

The diagnosis between spasmodic and inflammatory croup is as.
follows:

SPASMODIC CROUP.	INFLAMMATORY CROUP.
Onset sudden, usually at night, with few or no prodromal symptoms.	Onset gradual, with sore throat,. tickling, tenderness of larynx and catarrh.
Difficulty of swallowing absent or temporary.	Increasing diffiulty in swallow-- ing.
Febrile symptoms absent, or much less marked.	Flushed face, hot, dry skin, hight temperature (105°), frequent pulse..

SPASMODIC CROUP.	INFLAMMATORY CROUP.
Larynx little affected.	Mucous membrane of larynx red and swollen, sometimes œdematous.
Intermission complete, or nearly so, between the croupous attacks.	Remission but slight; local symptoms and pyrexia continue.
Very rarely fatal.	In early life a dangerous disease.

Very considerable differences of opinion are entertained as to the formidable and frequent disease *diphtheria*. Some maintain its identity with membranous croup, others with scarlatina, while others believe it to be a malady distinct in origin, course, result and treatment from them both. The last mentioned opinion appears to have the most adherents, and the most facts on its side. The differences between the diseases are fully set forth in the table subjoined:

MEMBRANOUS CROUP.	DIPHTHERIA.
It is a local complaint, rarely if ever occurring after puberty. A rare disease.	Is a general disease, common to all ages. A disease of frequent occurrence, often epidemic or endemic.
It is not contagious. Type sthenic.	It is decidedly contagious. Type asthenic.
Commences with a cough, catarrh and hoarseness; little or no sore throat and difficulty of swallowing. Cough shrill, metallic; breathing stridulous from the outset.	Commences with a chill, sore throat, difficulty of swallowing; but neither hoarseness nor cough at the outset. Stridulous breathing a late symptom.
The membranous affection begins in the larynx and extends to the throat.	The membranous affection begins in the throat and thence extends to the larynx (DA COSTA).
Fauces injected but rarely swollen, and generally without exudation.	Fauces injected, swollen and presenting exudation.
Exudation never cutaneous.	Exudation often cutaneous.
No swelling of the submaxillary glands.	Submaxillary glands swollen.
Epistaxis and albuminuria absent.	Epistaxis and albuminuria frequent.

MEMBRANOUS CROUP.	DIPHTHERIA.
Symptoms local; often no prostration of the general strength.	Considerable, often extreme prostration.
Relief follows the use of emetics, local counter-irritants, expectorants and depressants.	Demands a stimulating and sustaining treatment.
Is never followed by paralysis.	Subsequent paralysis not infrequent.
Less often fatal. Death from apnœa. Blood not changed. Spleen not affected.	Frequently fatal. Death usually by asthenia. Blood after death usually fluid and dirty brown. Spleen enlarged and softened (J. W. HOWARD).

With regard to pathology, a recent authority states that fibrinous and diphtheritic exudations are usually considered to be characteristic of croup and diphtheria, and yet diphtheritic inflammation is no more to be confounded with the disease diphtheria than is fibrinous inflammation with the disease croup (FITZ.*) Diphtheria may exist without the formation of false membrane, or various exudations may be present; mucous, fibrinous or diphtheritic. On the other hand, diphtheritic deposit occurs in other diseases than diphtheria, as in diphtheritic conjunctivitis, and diphtheritic inflammations of wounds and of variolous eruptions. The characteristics of a diphtheritic inflammation are the presence within the tissues of a clotted exudation, associated with defined swelling and local necrosis. With leucocytes, interlacing fibres and granular material, micro-organisms are detected by the microscope (*micrococci diphtheriae*).

TONSILLITIS.

Inflammation of the tonsils assumes two forms, in one of which, the *catarrhal* form, the inflammation extends to the secreting tissues and lining membrane of the crypts, and in the other to the parenchymatous structure of the gland. These two forms differ widely in cause, in symptoms, in treatment and result. Their diagnostic symptoms, as tabulated by Mr. ARTHUR TREHERNE NORTON,† are as follows:

---

* Vol. 1, Pepper's System of Medicine, page 50, Phila., 1885.

† *British Medical Journal*, Jan., 1874.

FOLLICULAR TONSILLITIS.	PARENCHYMATOUS TONSILLITIS.
Is a mucous inflammation of three or four days' duration.	Is a fibrous inflammation of from two to four weeks' duration.
Caused by exposure to draught, damp, cold, etc.	Often caused by neighboring inflammation, cutting wisdom teeth.
Prostration and often profuse perspiration. Pulse small and quick. Never runs on to abscess.	High fever, with hot, dry skin. Pulse strong and hard. Commonly forms an abscess.
Both tonsils affected.	Rarely both affected.
Lacunæ filled with masses of morbid secretion, may even resemble ulcers.	Often covered with lymph, but no collection of secretion in lacunæ.
No œdema around.	Extensive œdema.

DISEASES OF THE LUNGS.

In passing from the consideration of the disorders of the upper air-passages to the diseases of the lungs, it is thought advisable to discuss somewhat in detail the several methods of examination of the patient, and to consider systematically the various objective phenomena presented by them as introductory to the study of their alterations, which are characteristic of certain diseases. Palpation, mensuration, auscultation and percussion, therefore, furnish evidence of great clinical importance, which may be considered collectively under the head of physical diagnosis.

The study of Physical Diagnosis necessarily commences with a correct appreciation of the location of organs, their functions and physical characters in health; to which must follow a clear understanding of the specific and peculiar alterations which each of these elements undergoes when it becomes a factor in disease. To acquire this, we give on the following pages tabular arrangements of the following subjects:

I. The Regions of the Chest, their Contents and Normal Signs. II. The Normal Differences between the two Sides of the Chest. III. Methods of Physical Examination. IV. Normal Respiratory Sounds. V. Normal Voice Sounds. VI. Abnormal Resonance on Percussion, and its Causes. VII. Abnormal Intensity, Rhythm, and Quality of Respiratory Sounds. VIII. Abnormal (dry) Respiratory Sounds. IX. Abnormal (moist) Respiratory Sounds. X. Abnormal (amphoric) Respiratory Sounds. XI. Abnormal Voice Sounds.

I. THE REGIONS OF THE CHEST.

REGION.	CONTENTS.	RESONANCE ON PERCUSSION IN HEALTH.	AUSCULTATION IN HEALTH.
1. CERVICAL.	Larynx and trachea.		Tracheal breathing and voice.
2. SUPRA-CLAVICULAR.	Apex of lung.	Clear.	Very pure vesicular murmur (scarcely audible); voice scarcely audible.
3. CLAVICULAR.	Clavicles and vesicular structure of lung.	Clear.	Pure vesicular murmur and scarcely audible voice, except at the sternal end, w h e r e are bronchial breathing and bronchophony.
4. SUBCLAVIAN.	Vesicular structure of lung.	Clear. R a t h e r higher p i t c h e d percussion n o t e on right side than on the left.	Pure vesicular murmur and scarcely audible voice. Heart sounds on left side below.
5. MAMMARY.	Vesicular structure of lung. Heart on left side.	Clear on right side. Dull on left i n greater part of region.	Pure vesicular murmur above. Heart sounds below on left side, and feeble vesicular murmur on right. Voice scarcely audible.
6. INFRA-MAMMARY.	Anterior portion of base o f l u n g. Stomach below, on left side, liver on right.	Generally tympanitic on left side; dull on right.	Distinct vesicular murmur. Voice scarcely audible.
7. SUPERIOR STERNAL.	Division of t r a - chea, aorta, and great vessels.	Clear.	Bronchial breathing and bronchophony.
8. INFERIOR STERNAL.	Anterior mediastinum above. Stomach below.	Clear above; tympanitic below.	Pure vesicular murmur above, becoming feeble below. V o i c e scarcely audible.
9. AXILLARY.	Vesicular structure of lung.	Clear.	Pure vesicular murmur. Voice scarcely audible.
10. LATERAL.	Vesicular structure of lung.	Clear above; dull below on right side.	Pure vesicular murmur. Voice scarcely audible.
11. SUPRA-SCAPULAR.	Apex of lung.	Clear.	Pure vesicular murmur. Voice scarcely audible.
12. SCAPULAR.	Vesicular structure of lung.	Rather less clear.	Pure vesicular murmur. Voice scarcely audible.
13. INTER-SCAPULAR.	Roots of lung and large bronchi.	Clear.	Bronchial breathing and bronchophony.
14. INFRA-SCAPULAR.	Base of lung.	Clear.	Very pure v e s i c u l a r murmur. Voice scarcely audible.

II. NORMAL DIFFERENCES BETWEEN THE TWO SIDES OF THE CHEST. (A. H. SMITH.)

	RIGHT SIDE.	LEFT SIDE.
Percussion Resonance.		A little more intense than on the right side.
Vocal Resonance.	Decidedly greater on the right side.	
Bronchial Whisper.	More intense than on the left, and a little lower in pitch.	
Inspiratory Sound.		A little lower on this side, more vesicular in quality, and lower in pitch.
Expiration.	Frequently prolonged in healthy individuals on this side.	

III. METHODS OF PHYSICAL EXAMINATION.

METHODS OF EXAMINATION.	REVEALS.	INSTRUMENTS USED.
1. INSPECTION.	Form, symmetry and capacity of the chest. Local bulging, depression or retraction. Condition of intercostal spaces. Character and frequency of respiratory movements. Comparative size and degree of movement of the two sides. Position and extent of impulse of heart.	
2. PALPATION. (*Application of the hands.*)	Comparative movement of the two sides. Vibration communicated to the chest wall by the voice (vocal vibration or vocal fremitus.) Force of the heart's impulse. Occasionally certain morbid phenomena, as pleural and pericardial friction, valvular thrill.	
3. MENSURATION. (*a*) *Of Size.* (*b*) *Of Movement.*	Comparative size of the two sides of the chest. Actual and comparative movement of the chest in respiration.	Graduated tape. Cyrtometer. Dr. Sibson's stethometer. Dr. Quain's stethometer. Dr. Edward's chest calipers. Dr. Hutchinson's spirometer.
4. PERCUSSION.	Degree of resonance in various parts of the chest.	Plessor—A hammer tipped with india rubber.

III.—METHODS OF PHYSICAL EXAMINATION—(Continued.)

METHODS OF EXAMINATION.	REVEALS.	INSTRUMENTS USED.
	Extent of cardiac dulness.	The first and second fingers of the right hand will be found to be the best plessor. Pleximeter—A thin plate of ivory or bone. The forefinger of the left hand will be found to be the best pleximeter.
5. AUSCULTATION.	Character of respiratory murmur. Abnormal respiratory sounds. Heart sounds. Abnormal cardiac sounds.	Stethoscope—Made of wood, metal, or vulcanite. Dr. Scott Alison's bin-aural stethoscope.
6. SUCCUSSION.	Presence of air and fluid in pleural cavity.	

PERCUSSION may be—*Immediate*—Where the chest is struck *directly*, without the interposition of any pleximeter.

(2) *Mediate.*—Where, between the chest and the substance with which the stroke is made, an instrument termed a pleximeter is interposed. This may be either a thin plate of ivory or bone, or, still better, the first and second fingers of the left hand.

AUSCULTATION may be—*Immediate.*—Where the ear is applied *directly* to the walls of the chest.

(2) *Mediate.*—Where the stethoscope is interposed between the ear and the walls of the chest.

IV. NORMAL RESPIRATORY SOUNDS.

SOUND.	SITUATION WHERE HEARD.
VESICULAR BREATHING.	All over the chest except the upper part of the sternum and the space between the scapulæ, the respiratory sound being louder, and three or four times longer than the expiratory.
PUERILE BREATHING.	Is the loud vesicular breathing of children audible over the same parts of the chest as the ordinary vesicular breathing,

IV. NORMAL RESPIRATORY SOUNDS.—(*Continued.*)

SOUND.	SITUATION WHERE HEARD.
BRONCHIAL BREATHING.	Upper part of the sternum and the space between the scapulæ in many healthy persons.
TRACHEAL OR LARYNGEAL } BREATHING.	Over the trachea and larynx.

V. NORMAL VOICE SOUNDS.

SOUND.	SITUATION AND CHARACTER.
ORDINARY VOCAL RESONANCE.	Is the voice-sound heard over the pulmonary regions where vesicular murmur is audible. A muffled, diffused sound; the articulation of the voice is not appreciable.
NATURAL BRONCHOPHONY.	Heard over the upper part of the sternum, and between the scapulæ in a certain number of healthy persons. A more distinct and concentrated sound than the last variety.
LARYNGOPHONY AND TRACHOPHONY.	Voice-sounds heard over the larynx and trachea. Voice transmitted imperfectly articulated to the ear of the observer, with so much loudness and concentration as even to be painful.

VI. ABNORMAL RESONANCE ON PERCUSSION.

RESONANCE.	CAUSE.	EXAMPLES OF DISEASE.
DIMINISHED in various degrees or altogether ABSENT.	Deficiency of air, or abnormal deposit, in the lung beneath the part percussed; or solid or liquid matter between the walls of the chest and the lung containing air; or extreme distention of the chest with air.	Pneumonia, first stage. Phthisis; contracted lung, with thickened pleura. Œdema and congestion of lung. Tumors. Collapse of lung. Pneumonia, second and third stages. Intra-thoracic tumors and aneurisms. Effusions into pleural cavity, or its extreme distention by air.
INCREASED. TYMPANITIC.	Air increased in quantity, or air in pleural cavity.	Emphysema. Tubercular cavity, having thin walls, and situated near the surface. Pneumothorax. Extreme emphysema.
AMPHORIC. BOX-LIKE.	A large cavity (or conditions resembling it) with very tense walls, containing air.	Upper part of lung compressed by fluid below. Hydro-pneumothorax. Cavities.
CRACKED-POT SOUND.	Air expelled from cavity by sudden pressure.	Cavity of considerable size, with large bronchus opening into it, (mouth of patient being open during percussion.)

VII. ABNORMAL INTENSITY, RHYTHM, AND QUALITY OF RESPIRATORY SOUNDS.

	SOUNDS.	CHIEF CAUSES.	CONDITION OF ORGANS.	EXAMPLES OF DISEASE.
I. Changes in Intensity.	FEEBLE BREATHING.	Air entering the air cells in diminished quantity and force.	Lung partially solidified either by increase of solid or fluid within it, or by pressure from without; dilatation of the air-vesicles; in some cases lungs not affected.	Incipient phthisis. Bronchitis. Pneumonia, first stage. Tumors. Pleurisy. Emphysema. Pleurodynia.
	EXTINCT BREATHING.	The presence of a non-conducting medium between the lung and the chest-wall, or some impediment to the entrance of air into the bronchi.	Lung solidified by pressure upon its surface; plug of mucus, fibrinous exudation or foreign body in the bronchi, or tumor compressing the bronchi.	Pleuritic effusion. Pneumothorax. Plastic bronchitis. Tumors. Foreign body in bronchus.
	PUERILE, or SUPPLEMENTARY. } BREATHING.	Air entering the air-cells with increased rapidity and force.	Healthy, but exaggerated in function.	Disease of opposite lung or of other parts of the same lung. Met with as a normal condition in childhood.
II. Changes in Rhythm.	INTERRUPTED, JERKING, COGGED-WHEEL } BREATHING.	Respiratory movements restrained by pain, or mental emotion, or some temporary local obstruction of the air-tubes.	Varies with the disease causing it.	Pleurodynia. Pleurisy. Debility, with palpitation. Hysteria. Incipient phthisis. Spasmodic asthma.
	PROLONGED EXPIRATION.	Loss of elasticity in the lung tissue.	Thinning of the walls of the air vesicles, with dilatation and destruction of the alveolar septa.	Emphysema.

ABNORMAL INTENSITY, RHYTHM AND QUALITY OF RESPIRATORY SOUNDS—(*Continued.*).

SOUNDS.		CHIEF CAUSES.	CONDITION OF ORGANS.	EXAMPLES OF DISEASE.
III. Changes in Quality.	EXAGGER- ATED, COARSE } BREATHING.	Increased friction in the air-cells and smaller bronchial tubes.	Lung not solidi- fied (soft sound).	Generally con- sistent with health and supplement- ary. Heard in cases of uræmia and other blood poisoned dis- eases, and in hysteria and nervous dis- eases.
			Lung solidifie̓d or bronchial tubes obstructed (harsh sound).	Incipient phthisi̓s.
	BLOWING, TUBULAR, OR BRON- CHIAL CA- VERNOUS } BREATHING.	Friction of air in the bronchial tubes, or in cav- ities of the lung.	Condensation of the lung between the chest wall and the larger bronchi or cavities.	Phthisis. Pneumonia. Tumors. Tubercular and other cavities.
	AMPHORIC BREATHING.	Air passing into a large cavity with dense walls.	Cavities with dense walls.	Pneumothorax. Dilated bron- chi. Large cavities.

VIII. ABNORMAL DRY RESPIRATORY SOUNDS.

SOUND.	SITUATION.	CAUSE.	EXAMPLES OF DISEASE.
SIBILUS.	Smaller bronchial tubes.	Vibration of thick mucus attached to the wall of the tube, or contrac- tion of the tube, due either to swel- ling or spasm ; not easily removed by cough.	Bronchitis. Emphysema. Asthma.
RHONCHUS.	Larger bronchial tubes.	Vibration of thick mucus in tubes ; generally easily removed by cough.	Bronchitis.

CLICKING OR CRACKLING.

DRY CRACKLING.	Smaller Bronchi.	Separation of the ad- herent walls of the bronchi—the dry tending to pass into the moist variety.	Incipient phthisis.
HUMID CRACKLING.	Smaller bronchi.		Softening tubercle.
PLEURAL FRICTION SOUND.	} Layers of pleura.	Movement of opposed surfaces of pleura roughened by the de- posit of lymph or tu- bercle.	Pleurisy before effu- sion has commenced, or after absorption of the fluid.
CREAKING SOUND.			

11

IX. ABNORMAL MOIST RESPIRATORY SOUNDS.

SOUND.	SITUATION.	CAUSE.	EXAMPLES OF DISEASE.
CREPITANT RALE. (*Fine or pneumonic crepitation*).	Air-vesicles.	Opening up of collapsed air-cells, or separation of their adherent walls.	Pneumonia in first stage. Œdema of lungs. Collapse.
SUBCREPITANT RALE. (*Medium crepitation*).	Smaller bronchial tubes.	Bursting of air-bubbles in fluid.	Capillary bronchitis. Phthisical bronchitis. Resolution of pneumonia. Œdema of lung. Pulmonary apoplexy.
MUCOUS RALE. (*Large crepitation*.)	Larger tubes and small or moderate-sized cavities.	Bursting of air-bubbles in fluid.	Phthisis. Bronchitis. Hæmoptysis.
GURGLING OR CAVERNOUS RALE.	Large cavities (or number of small cavities).	Bursting of air-bubbles in fluid.	Phthisis (3d stage). Bronchiectasis.
CHURNING SOUND.	Lung in a state of disorganization.		Abscess of lungs. Gangrene of lung.

X. ABNORMAL AMPHORIC SOUND.

SOUND.	SITUATION.	CAUSE.	EXAMPLES OF DISEASE.
SPLASH ON SUCCUSSION.	Cavity of pleura or large cavity.	Sudden disturbance of air and fluid existing together.	Pneumothorax with effusion. Very large cavity.
BELL SOUND.	Cavity of pleura.	Auscultation of an air-containing cavity while an assistant uses two coins, one as a hammer, the other as a pleximeter.	Pneumothorax.
AMPHORIC ECHO AND METALLIC TINKLING.	Cavities.	Vibration of air in large cavities with tense walls. The former may be produced by râles and rhonchi in the chest, by the voice, and by the act of coughing; the latter requires, in addition, a little fluid at the bottom of the cavity, set in vibration by a momentary impulse, such as the fall of a drop of fluid, and is essentially the echo of a bubble.	Phthisis with very large cavities. Pneumothorax with effusion.

XI. ABNORMAL VOICE SOUNDS.

SOUND OF VOICE.	CHARACTER OF SOUND.	CAUSE.	EXAMPLES OF DISEASE.
FEEBLE OR ABSENT VOCAL RESONANCE.	The obscure humming of buzzing noise heard over the normal chest either very feeble or altogether absent.	Primary bronchus obstructed; conducting medium in pleura or rarefied condition of lung.	Tumors compressing, or foreign body in bronchus. Pneumothorax. Pleuritic effusion. Emphysema.
EXAGGERATED VOCAL RESONANCE.	Voice sounds unaltered in quality or distribution, but louder and of greater intensity than natural.	Increased resounding or conducting power due to consolidation of the lung, or to the formation of abnormal spaces.	Incipient phthisis. Dilatation of bronchi.
BRONCHOPHONY.	Voice-sounds heard louder, clearer, and more vibratory than natural, but unattended with articulation or tactile sensation to the ear.	Much increased resounding or conducting power.	Cavities due to phthisis or dilatation of the bronchi. Consolidation of the lung resulting from collapse, hæmorrhagic infarctions, pneumonia, phthisis, cancer, etc.
PECTORILOQUY.	Voice-sounds distinctly articulated and concentrated and as if spoken into the end of the stethoscope.	Large abnormal cavity with dense walls.	Phthisis, dilated bronchi, etc.
AMPHORIC RESONANCE OR ECHO.	A ringing metallic sound resembling that produced by speaking into an empty jar.	The voice reverberating in a large cavity with a small aperture.	Phthisis. Pneumothorax,
ŒGOPHONY.	A tremulous vibratory sound resembling the bleating of a goat, or the nasal Punchinello voice.	A thin layer of fluid in the pleural cavity, with condensed lung behind.	Pleurisy with effusion.

The quality and pitch of the vocal resonance varies greatly in different individuals, and as a diagnostic aid is almost useless in women. Whispering, sometimes, will give more satisfactory results than the loud " one, two, three," that is so constantly heard in our clinical amphitheatres. The use of a small reed whistle, such as is found in children's rubber toys, will often give more uniform effects for comparison than the voice.

GENERAL RULES FOR THE DIAGNOSIS OF DISEASES OF THE RESPIRATORY SYSTEM.

The late Dr. JOHN HUGHES BENNETT laid down the following practical rules:

1. A friction murmur heard over the pulmonary organs indicates a pleuritic exudation.

2. Moist or dry râles, without dulness on percussion, or increased vocal resonance, indicate bronchitis.

3. Dry râles accompanying expiration, with unusual resonance on percussion, indicate emphysema.

4. A moist râle at the base of the lung, with dulness on percussion, and increased vocal resonance, indicates pneumonia.

5. Harshness of the respiratory murmur, prolonged expiration and increased vocal resonance confined to the apex of the lung, indicates incipient phthisis.

6. Moist râles with dulness on percussion, and increased vocal resonance at the apex of the lung, indicate either advanced phthisis or pneumonia, generally phthisis.

7. Circumscribed bronchophony or pectoriloquy, with cavernous dry or moist râles, indicates a cavity. This may be one dependent on tubercular ulceration, a gangrenous abscess, or a bronchial dilatation. The first is generally at the apex, and the last two about the centre of the lung.

8. Total absence of respiration indicates a collection of fluid or of air in the pleural cavity. In the former case there is diffused dulness, and in the latter diffused tympany on percussion.

9. Marked permanent dulness, with increased vocal resonance and diminution or absence of respiration, may depend on a chronic plastic pleurisy, a thoracic aneurism, or a cancerous tumor of the lung.

THE FORMS OF PHTHISIS.

Most systematic writers, both in the United States and Europe, are agreed in recognizing three principal clinical varieties of phthisis. It is of import, both to the prognosis and therapeutics of the case, to distinguish these aspects of the disease; and although in many cases the type is by no means prominently defined, in the majority there is no great difficulty in assigning them to one or another class. The three forms are:

1. Catarrh'al or inflammatory phthisis: " Desquamative pneumonic phthisis." (BUHL.) Chronic broncho-pneumonia.

2. Fibroid phthisis. Cirrhosis of the lung. Chronic pneumonic phthisis. Bronchial phthisis. Chronic interstitial pneumonia.

3. True primary tuberculosis. Tubercular phthisis. Tubercular pneumonia. (DA COSTA.)

On the clinical recognition of these three varieties, Dr. ALFRED L. LOOMIS says :

If a case of phthisis present himself for examination, and it is found that the disease began with the ordinary symptoms of a cold, and that this cold periodically improved and relapsed, but that the cough never left him, but became more pronounced and deepened into what we usually find in advanced phthisis, accompanied with emaciation and occasional hæmoptysis, we are in a position to say that the patient presents the usual characteristics of *catarrhal* phthisis.

If, however, he gives a history of persistent cough for many years, as is found in chronic bronchitis, and eventually furnishes the rational history of advanced phthisis, with the presence of cavities in the lung, we may consider him as having the disease of the *fibrous* form, in which cavities are the result of dilated bronchi.

Finally, if the patient says that an early symptom was emaciation, with impaired digestion, accompanied by a dry, hacking cough, and if, moreover, there was a steady rise in the temperature, we are justified in suspecting the presence of *tubercular* phthisis.

Prof. AUSTIN FLINT* contended that there is no pathological distinction between the catarrhal and tubercular form of phthisis. The rare form of pulmonary phthisis, characterized by great predominance of interstitial growth, leading to notable diminution of the volume of lung by atrophy and dilatation of bronchial tubes, may be conveniently considered under a separate heading, as "fibroid" phthisis. Acute

* Pepper's System of Medicine, Vol. iii., p. 391.

miliary tuberculosis may complicate either of these forms, or occur as a primary disease running a rapid course.

THE FORMS OF PHTHISIS—CHRONIC CATARRHAL PNEUMONIA.

PERIOD OF INVASION . | Precursory catarrh, sometimes pneumonia, croup, measles, or other inflammatory disease; cough "deepens," proceeding from the trachea to the alveoli and bronchioles, indicated by dark yellow streaks in the sputum. Fever and wasting not marked at outset. Hæmoptysis not common at this period.

TEMPERATURE . . . . The hectic is more of a *remittent* or *intermittent* than of a continued type; with a range of, say, 1.1° C. between evening and morning temperature; the evening elevation being a constant feature.

The fever may present all possible variations in the same individual. A sudden accession may be regared as an indication of some fresh inflammatory process; *e. g.*, pleuritis, pneumonia.

With marked evening rise of temperature, the rate of respiration does not correspondingly accelerate; hardly ever more than six or eight breaths per minute.

PHYSICAL SIGNS . . . | In the first stage, feeble, harsh or puerile respiratory sounds are heard, with all the signs of catarrh at apices and elsewhere.

Dulness usually marked; when its area accords with the other signs it is a comparatively favorable feature.

The presence of lobular infiltration may, in some cases, cause a hollow or tympanitic note.

"Cracked-pot" resonance over a cavity with thin walls.

Fremitus is intensified over cavities connecting with bronchi and containing air.

Bronchial respiration, bronchophony, and sonorous râles are heard after extensive induration.

THE FORMS OF PHTHISIS—CHRONIC CATARRHAL PNEUMONIA.—*Continued.*

GENERAL NUTRITION. | Not impaired in the early stage, but when cavities form, hectic and emaciation set in and we have " pneumonic phthisis."

May continue for years, until pneumonic phthisis is developed, when it lasts only a few months.

COMPARISON OF THE FORMS OF PHTHISIS.

FIBROID.	TUBERCULAR.
More or less dyspnœa, gradually increasing. Cough worse in winter, sometimes absent in summer. Hæmoptysis frequent. Pulse slightly rapid, perhaps irregular. Expectoration often profuse, mucous or muco-purulent. No bacilli in expectorated matters.	Commences in the alveoli, bronchioles, or connective tissue. Pallor, fever, emaciation and night-sweats early. Cough hoarse and hard, voice hoarse or inaudible, distressing laryngitis. The sputa reiain the crude character of the mucous sputa of acute bronchitis, Bacillus tuberculosis (KOCH) found in sputum.* Spleen somewhat enlarged.
Elevation of temperature and other febrile symptoms very variable, sometimes wholly absent (BRISTOWE). No special type.	The hectic is of a *continued* type; temperature always above normal, but not much higher in the evening than in the morning; *i. e.*, the remissions *not* well marked; moreover, it resists treatment.
Notable dulness on percussion, resonance sometimes tympanitic. Respiration bronchial, or bronchovesicular. Bronchophony and increased vocal resonance. The affected side becomes contracted either entirely or in part.	Signs not well marked, not sufficiently so to account for the symptoms. Solidification not extensive. Expansion unequal.
Bronchial dilatation (fusiform) gives the physical sign of a cavity.	Cavities form after softening with destruction of lung tissue.
Not incompatible with apparent good health.	Health obviously impaired.
Duration indefinite.	Lasts about one year.

* For method of detecting the bacillus tuberculosis, see page 65.

THE DIAGNOSIS OF INCIPIENT PHTHISIS.

With the exception of the detection of the tubercle bacilli, there is no absolutely sure symptom of phthisis previous to percussion dullness, but a very strong presumption of its approach can be drawn from the presence of the following physical changes :

1. *Emaciation.* Where there is progressive emaciation without assignable cause, and especially if the appetite continue good, phthisis should always be suspected. The loss of flesh first shows itself in a retraction of the skin over the cheeks, a thinning of the lips and ears, and a pinched appearance of the nose. The nostril on the affected side is usually slightly more dilated than the other.

2. *Anæmia*, seen in the bluish hue of the sclerotic, and in the pallor of the cheeks.

3. *Sore throat* and hoarseness. A very early symptom. On examination the pillars of the fauces are found hyperæmic, the throat congested and the bronchial glands enlarged.

4. *Swelling of mucous membrane of larynx*, especially forming a turban-shaped epiglottis, which at the same time assumes a horse-shoe bend ; and pyriform enlargement over the arytenoid cartilages (SEILER).*

5. *Depression of the acromial end of the clavicle*, on the affected side. In health the acromial end is slightly higher than the sternal end.

6. *Rheumatoid pains* in the arms coming suddenly at night or in the early morning, not increased on moving the arm.

7. *Pityriasis versicolor*, in the form of pale yellow or reddish spots appearing on the skin of the chest, neck and arms. This is considered by AUFRECHT a very characteristic symptom.

8. In regard to the *breathing*, what is considered as suspicious are weak, jerking, " cogged-wheel," or sonorous sounds, rough breathing, a lengthened strong expiration after soft inspiration, especially when in circumscribed regions these sounds differ from those on the other side of the chest. The most appropriate spot to note the duration of expiration is over the larynx or trachea. In proportion as the tubercular deposit is more extended, the expiratory murmur becomes more tubal in quality and higher in pitch (ARMOR). In the normal chest, the respiratory sound becomes weaker in the supra-spinous region

---

* Proceedings Philadelphia Co. Medical Society, Vol. ii, p. 101, Philada., 1880.

outward from the vertical column. Dr. HEITLER considers it, therefore, strong evidence of incipient pulmonary phthisis if the respiratory sounds during expiration are more sonorous over these regions than near to the vertebral column.*

9. *Unequal expansion of chest* is an early sign of commencing disease of the apex. The expansion is less on the diseased side.

10. *Alterations in temperature curve* frequently take place early. The temperature may be low, but the characteristic range will be : (1) a marked rise after 2 p. m. ; (2) a rapid fall after 10 p. m. ; (3) minimum about 7 a. m.; (4) recovery to normal about 10 a. m. (C. T. WILLIAMS). Such a curve must always excite grave suspicions.

11. *Rapidity of pulse.* A persistent and sustained increase in the pulse rate, without cardiac disease, is a valuable rational sign, present very early in most cases.

12. The *cough* of incipient phthisis is usually short, hacking, and dry, or with a slight, glairy, mucous expectoration only. From the presence of fragments of the pulmonary fibrous tissue in the sputum, "we are sometimes enabled to suspect the existence of consumption before the physical signs of even its early stages are well defined." (DA COSTA.)†

13. *Hæmoptysis.* The appearance of hæmoptysis is always a serious element of diagnosis. Light, frothy, red blood, rising without apparent exertion, is an indication which, in America at least, has proved of graver meaning the more it has been investigated.‡ On the other hand, cases will be met with sometimes, in whom there may be considerable hæmoptysis, with marked dulness at the apex, without association with tubercle.§

---

* DOBELL'S Reports on Diseases of the Chest, 1877.

† To examine sputa for elastic fibres, mix it with a soda solution :

    ℞.   Liquor sodæ,   1 part.

           Aquæ destill.,   2 parts.   M.

And boil for four or five minutes. Then dilute with an equal quantity of distilled water, and pour into a flat porcelain vessel. The particles suspended in the water may then be taken out and examined under the microscope. The fibres in this process are brown, slightly reticulated, and a fraction of a millimetre in length (SOKOLOWSKI).

‡ See second Report of the New York Mutual Life Insurance Company, 1877.

§ See Prof. DA COSTA, in *Medical and Surgical Reporter*, July 13, 1878.

14. *Clubbing of the finger ends,* when associated with incurvation of the sides and tips of the nails, means obstruction of the subclavian veins, which is one of the earliest effects of tuberculosis; but clubbing without this incurvation is rather against the probability of tubercle (DOBELL).

15. *Amennorrhœa* is, in young females, often one of the earliest signs of phthisis.

16. A *red line* is occasionally noticed on the gums at the base of the teeth.

17. *Arthritis.* M. LAVERAN* has drawn attention to the occasional occurrence of arthritis as the first symptom of a general tuberculosis.

DIAGNOSIS BETWEEN INCIPIENT PHTHISIS AND BRONCHITIS.

INCIPIENT PHTHISIS.	BRONCHITIS.
1. The cough commences gradually, without marked disturbance or coryza, often preceded by slight loss of flesh and strength.	1. The cough commences suddenly, and is usually ushered in by feverishness and coryza.
2. The cough is generally dry and hacking at commencement, followed by the expectoration of a thin mucous fluid, which soon becomes thick and opaque, or is slightly streaked with blood.	2. The cough is accompanied with expectoration almost from the first; generally abundant; frothy or muco-purulent; not rarely blood-stained.
3. Examination by the microscope shows portions of lung tissue (yellow elastic fibres) in the sputa.	3. No evidence of destruction of lung tissue.
4. Pain of a wandering character about the chest, especially under the clavicles or between the shoulders.	4. A feeling of tightness and rawness behind the sternum, aggravated by coughing.
5. Evening rise of temperature.	5. Elevation of temperature not particularly marked toward evening.

* *Le Progrès Médical,* October 25, 1876. Quoted by Dr. M. ANDERSON.

DIAGNOSIS BETWEEN INCIPIENT PHTHISIS AND BRONCHITIS—(*Continued.*)

INCIPIENT PHTHISIS.	BRONCHITIS.
6. The morbid physical signs usually confined to upper lobe of one side; are very persistent, and if on both sides at first, apt to subside on one and increase on the other.	6. Morbid signs usually predominating in the lower lobes; are on both sides; are of temporary duration, and subside gradually and equally on both sides.
7. Family history and general appearance indicate tuberculous cachexia. Most frequent in youth.	7. No marked hereditary tendency; common at all ages.
8. Essentially chronic.	8. Has an acute beginning.
9. Bacillus tuberculosis occurs in sputa.	9. No bacilli in sputa.

While these points of difference between tubercular disease and catarrhal inflammation of the mucous membrane lining the bronchial tubes are in the main reliable, yet it must not be forgotten that chronic bronchitis is often attended by structural changes in the lung, leading in one set of cases to increase of connective tissue, with dilated bronchiæ—fibroid degeneration, chronic broncho-pneumonia—and in another to deposits, chiefly epithelial, in the air-cells, producing spots of consolidation.

The general clinical history of the three stages of phthisis may be summarized in the following brief table:

PULMONARY PHTHISIS. (CHRONIC TUBERCULAR PNEUMONIA.)

STAGE OF DISEASE.	SYMPTOMS.	PHYSICAL SIGNS.
1st stage (incipient). Stage of invasion.	Cough at first dry, then with expectoration of mucus, frequently streaked or dotted with blood, or with copious hæmoptysis. Dyspnœa. Pains in the various parts of the chest, especially on the affected side. Dislike to fatty articles, and other dyspeptic symptoms; tendency to vomiting after paroxysms of coughing. Night-sweats. Emaciation. In females, disturbance of the catamenial functions. Occasionally hectic.	Diminished movements. Increased vocal fremitus. Loss of percussion resonance, rise in pitch, or a boxy, wooden note beneath the clavicle or in the interscapular region. Feeble, coarse, or interrupted vesicular murmur, with prolonged expiration. Increased vocal resonance. Occasional sibilus or creaking friction sound. Heart sounds abnormally loud over affected side. Subclavian murmur. Puerile (exaggerated) respiration on sound side.
2d stage (confirmed). Stage of deposit	Cough more severe, with puriform expectoration, of a yellow or greenish hue, and often bloody. Profuse night-sweats and rapidly progressive emaciation. Pinched and anxious expression. Loss of appetite. Thirst. Diarrhœa. Sometimes hectic.	Greater diminution of movement of the affected side, and some amount of flattening. Increased vocal fremitus. Increased dulness, extending downward. Bronchial breathing, mixed with mucous râles or with click at the end of each inspiration. Bronchophony.
3d stage (advanced). Stage of softening and formation of cavities.	Cough rather looser, still with puriform (nummular) expectoration, or attacks of copious hæmoptysis. Extreme emaciation and debility, with or without night-sweats. Voice husky and hollow. Aphthæ on mouth and fauces. Hectic. Clubbed fingers and talon-like nails.	Scarcely any movement of the affected side. Marked flattening. Increased vocal fremitus. Dulness less marked. Box-like resonance or cracked-pot sound. Cavernous breathing, with gurgling and splash on cough. Occasionally metallic sounds. Pectoriloquy.

PHTHISIS—(*Continued*).

COMPLICATIONS NOT RESTRICTED TO ANY PARTICULAR STAGE OF PHTHISIS.

The chief of these are : Affections of the larynx and trachea, especially ulceration; bronchitis, intercurrent pneumonia, or pleurisy; perforation of the pleura, with pneumo-hydrothorax or empyema ; enlargement of the external absorbent glands, or of those in the chest and abdomen ; tubercular peritonitis; ulceration of the intestines, especially the ileum; fatty or amyloid liver; fistula in ano; various forms of Bright's disease; diabetes; pyelitis; tubercular meningitis, or tubercle in the brain, and thrombosis of the veins of the legs.

POST-MORTEM APPEARANCES.

First stage. Lesions most marked at, or even confined to, one apex, where are to be seen gray, semi-transparent nodules, varying in size from a small pin's head to a hemp-seed; the lung-tissue around these nodules may be healthy, but is generally hyperæmic and congested, slightly increased in density. In more advanced cases, in addition to the miliary nodules, there may be small yellow masses, less defined, but larger than the gray variety. Both kinds may either be scattered or several in one group, forming a considerable mass.

Second stage. Commencement of caseation and softening in the centre of the consolidated portions, inflammation of the surrounding parenchyma, together with obliteration of the blood-vessels and formation of cicatricial tissue.

Third stage. Cavities of various sizes and forms, and either single or numerous, generally containing puriform fluid. Ulceration and dilation of the bronchial tubes. Lung indurated and puckered in proportion to chronicity of disease.

ACUTE PHTHISIS,* ACUTE MILIARY TUBERCULOSIS, GALLOPING CONSUMPTION.

The formidable disease known under these names is probably, as M. BOUCHUT remarks, more common than is generally supposed, as it is generally mistaken either for capillary bronchitis or typhoid fever,

---

* Austin Flint criticises this application of the word, since phthisis means an essentially chronic process.

especially the latter. Its duration is brief, sometimes less than a fort-
night (DA COSTA), and its termination almost invariably fatal. Its
features are thus so entirely distinct from the chronic form, from the
clinical point of view, as to really constitute it a separate disease.

Its onset is marked by chills and feverishness, nausea, vomiting and
diarrhœa. There is a rapid pulse; dyspnœa; slight pain in the chest;
cough, usually with profuse expectoration. Great exhaustion, sweats,
rapid emaciation, and delirium, soon follow. One on both lungs exhibit unusual dullness, while the auscultatory sounds differ greatly in
different cases. The symptoms often simulate typhoid fever, so as to
distract attention from the pulmonary lesion.

The following are the marked diagnostic features of the disease:

1. *Facial expression.* The countenance is livid, indicating plainly
an impediment to the passage of blood through the lungs. In severe
typhoid fever the cheeks are slightly flushed, the facial muscles trem-
ulous, the eyes dull, and the mouth partly opened, presenting an appearance characteristic of the disease.*

2. *The delirium* of acute phthisis is restless and often violent, but
the rambling and wild talk is connected usually with things present
or near. In typhoid fever the delirium is generally muttering and low;
the mind deals with things absent, and the patient "is like a man talk-
ing in his dreams." (WATSON.)

3. *The tongue* in acute phthisis, at first covered with a white fur,
soon becomes red, glassy and dry. In typhoid it usually changes to
a brownish hue.

4. *The ophthalmoscope* is a most positive aid to the diagnosis, according to M. BOUCHUT. In all cases of acute, general, miliary tuberculosis, an ophthalmoscopic examination will reveal the presence of
tubercular granulations in the choroid,† thus placing the nature of the
disease beyond doubt.

5. *Abdominal symptoms.* Diarrhœa and gastric and abdominal pains
are often present in acute phthisis; but the red spots of typhoid are
not seen.

6. *Chest symptoms.* Dyspnœa is present always, but the orthopnœa

---

* L. J. WOOLLEN, *American Practitioner*, July, 1871.

† *Medical Times and Gazette*, January, 1875.

of capillary bronchitis is rare. (SHAW.) The respiration is greatly quickened, and the proportion to the pulse averages 1:3. (WALSHE.) The presence of percussion dulness, a sinking in at the upper part of the ehest, and the occurrence of hemorrhage, are conclusive evidence of tubercle. (DA COSTA.)

7. The sputum shows the characteristic serous and muco-purulent character, and may contain the elastic fibres of lung tissue, and numerous bacilli.

DIAGNOSIS OF SYPHILITIC PHTHISIS.

The distinctive traits of this form of lung disease have been separately studied by Dr. MACSWINNEY, of Dublin, and Dr. PENTIMALLI, of Naples. From these and a recent article on syphilitic diseases of the lung by Dr. E. T. BRUEN, of Philadelphia, in Pepper's System of Medicine, we derive the following scheme:

1. Absence of hereditary tendency, of a phthisical habitus, and of preceding pulmonary affections. Phthisis may complicate syphilitic cachexia.

2. History of syphilitic disease in other organs, and presence of the syphilitic cachexia in its tertiary stage. Substernal tenderness, thickening of the tibial periosteum or that of the head of one or of both clavicles.

3. The disease never begins in the apex, and is limited in its seat, being unilateral and generally posterior. (PENTIMALLI.) The tendency to localization in portions of the lungs, leaving large areas free from disease, is of value in diagnosis. (BRUEN.)

4. Hæmoptysis rare, febrile symptoms absent or slight.

5. Slowness in development, the acuter phthsical symptoms not manifest. Emaciation less rapid than in phthisis.

6. Exacerbations of pain during the night.

7. A peculiarly fetid breath may be noticed.

8. Reference of the feeling of oppression to the larynx rather than to the chest.

9. Failure of ordinary measures, and improvement under specific medication.

BRONCHITIS, ACUTE AND CHRONIC.

In most cases of bronchitis the inflammation is seated in the larger bronchial tubes. There is more or less swelling of their lining mucous membrane, not generally sufficient to prevent a free passage to the tidal air. The characters of the acute and chronic form are set forth in the tables, in the following pages.

There is a variety of chronic bronchitis, in which the material exuded on the surface of the air passages contains a large proportion of a fibrinous constituent which makes it tough and consistent, so that when expelled the substance appears as a perfect cast of the bronchial tube in which it was formed. This is called fibrinous bronchitis, and does not differ in pathology from the ordinary chronic variety, but is less susceptible to treatment. Its diagnosis is made from the appearance of the casts, and needs no further mention here.

ACUTE BRONCHITIS.

	SYMPTOMS.	PHYSICAL SIGNS.
1st or Dry Stage.	Chilliness, followed by frequent pulse and febrile symptoms; pains in limbs. Substernal pain. Hoarse dry cough. Feeling of oppression and tightness about the chest.	Breathing hurried. Rhonchal fremitus may be felt. Resonance on percussion unimpaired. Feeble vesicular murmur, mixed with rhonchus and sibilus. Puerile breathing in unobstructed parts of lung. Vocal resonance not materially altered.
2d or Moist Stage.	Cough, with expectoration of frothy, transparent mucus, mixed with air-bubbles of various sizes, and occasionally tinged or streaked with blood. Urgent dyspnœa, often amounting to orthopnœa. Lividity and febrile symptoms increased. Restlessness at night.	Breathing hurried. Rhonchal fremitus may be felt. Resonance on percussion clear, or only very slightly impaired. Feeble vesicular murmur mixed with rhonchus, sibilus and mucous râles. Vocal resonance unaltered.

ACUTE BRONCHITIS.

	SYMPTOMS.	PHYSICAL SIGNS.
3d Stage. a. (Termination favorable.)	Gradual remission of the symptoms. Expectoration becomes thick, greenish, and opaque, and sometimes nummulated.	Less amount of sonoro-sibilant and mucous râles, with return of normal vesicular breathing.
b. (Unfavorable.)	Dyspnœa very urgent, signs of impending suffocation. Profuse cold sweats. Sinking, drowsiness and delirium. Less cough, absence of expectoration.	In addition to the signs of the second stage, tracheal râles may be heard.

The *post-mortem* appearances are: Congestion of mucous membrane of bronchial tubes, with some degree of swelling and dryness of surface.

Lungs do not collapse when the chest is opened; nor do sections sink in water. The mucous membrane of the bronchi is red and swollen, and the tubes filled with frothy, adhesive mucus.

CHRONIC BRONCHITIS.

SYMPTOMS.

Two chief forms: the one characterized by the sputa being expectorated with great difficulty, consisting of small, gray, semitransparent pellets, and tending toward emphysema; in the other the sputa are abundant, muco-purulent, and brought up with ease; dilatation of the bronchi frequently associated with this form. The cough generally comes on at the approach of winter, with history of former attacks. Dyspnœa; lividity of surface; and in some cases the symptoms resemble those of chronic phthisis, as wasting, with night-sweats and hectic.

PHYSICAL SIGNS.

Respiration labored and abdominal. Vocal fremitus not materially altered; rhonchal fremitus can generally be felt. Impairment of resonance or a hyper-resonant note, according as collapse of lung and consolidation, or emphysema predominate, the former most marked at the bases, the latter at the anterior part. Feeble vesicular murmur. Rhonchus, sibilus, and mucous râles. Vocal resonance varies.

12

The *post-mortem* appearances are :—

Lungs generally much congested, presenting a dark livid hue, with portions collapsed, and others emphysematous. Bronchial tubes frequently dilated. Mucous membrane thickened, uneven, sometimes ulcerated, covered by a thick, puriform secretion, or sparingly coated by a tenacious, glairy, semi-transparent substance.

The principal diseases with which bronchitis may be confounded are pneumonia, pleurisy, and pulmonary phthisis. But each of these is characterized by the presence of definite physical signs, which are not to be found in ordinary bronchitis. For instance, in this disease there is no disparity between the two sides of the chest in the resonance obtained by percussion, nor in vocal resonance, the bronchial whisper and fremitus. The swelling of the bronchial mucous membrane may cause some diminution of the intensity of the vesicular murmur; but as the affection is bilateral and the bronchial tubes on both sides are affected equally both in degree and extent, there is no appreciable disparity between the two sides. Sometimes temporary weakening or suppression of the murmur may be caused by a plug of mucus, which will be detected on a second examination (FLINT), or by instructing the patient to cough, so as to dislodge it.

CAPILLARY BRONCHITIS.

Acute capillary bronchitis may, however, be taken for some of the forms of pneumonia, and in fact the descriptions of some writers would lead to the belief that they have committed this error. The following distinctions will make the diagnosis easy in most cases :—

CAPPILLARY BRONCHITIS.	PNEUMONIA.
Commences in the external air passages as a common cold and extends downward. Usually preceded by ordinary bronchitis.	Commences suddenly with a chill, and attacks the lungs directly. No antecedent bronchitis as a rule.
Always bilateral.	Generally unilateral.
Normal or exaggerated resonance on percussion, unless collapse has commenced.	Dulness on percussion more or less extensive at the outset.
Sub-crepitant râles on both sides of the chest.	Crepitant râle over affected lung tissue.

CAPILLARY BRONCHITIS.

Respiration not bronchial, 50 or more; pulse 150 or more.

Muco-purulent expectoration; no plastic lymph.

Dyspnœa intense; cyanosis early. No pain or but little.

Death from asphyxia; mortality more than half.

A disease of children.

PNEUMONIA.

Respiration bronchial, 25 to 40 per minute. Pulse 100 to 130.

Rust-colored expectoration; plastic lymph in pulmonary air cells.

Dyspnœa less; cyanosis late if at all. Pain in the side.

Death from asthenia; mortality ten per cent.

A disease of adult life.

PNEUMONIA AND PLEURISY.

Ordinary acute inflammation of the lungs in its early or first stage is well marked by the presence of a moderate or slight dulness on percussion over the affected lobe, and the detection on auscultation of *the crepitant* râle. The latter is indeed not invariably present, but when it is, taken in connection with the symptoms, it is pathognomonic.

Later in the disease the rust-colored expectoration of pneumonia on the one hand, and the physical signs of effused liquid into the pleural cavity in pleurisy on the other hand, offer distinctive features.

The general clinical histories of the diseases are given in the following tables:—

PNEUMONIA.

	SYMPTOMS.	PHYSICAL SIGNS.
1st Stage. (Engorgement.)	Single, severe rigor (or convulsions in children), followed by heat of skin. Increased frequency of pulse. Respiration greatly accelerated, with consequent disturbance of the pulse-respiration ratio. Dyspnœa. Pain in the side, increased by cough or deep inspiration. Cough, at first dry, with rusty sputa about the second or third day. Inability to lie on affected side. Dilated alæ nasi. Herpes about lips.	Diminished movement on the affected side. Respiration abdominal. Vocal fremitus normal. Percussion note not materially affected. Feeble vesicular breathing. Fine crepitant râles, most frequently heard at base of lung and at the end of inspiration. Presence of pneumococcus in the sputa.

PNEUMONIA—(*Continued.*)

	SYMPTOMS.	PHYSICAL SIGNS.
2d Stage. (Red hepatization.)	Increased distress and dyspnœa. Respiration and speech panting. Cough more urgent, and sputa still rust-colored, extremely viscid, and tenacious. Absence or deficiency of chlorides in the urine.	Very slight movement. Vocal vibrations well marked. Dulness on percussion. Tubular breathing and bronchophony, generally accompanied by some râles, if at the commencement of the 2d stage of a crepitant character, and afterward of a mucous nature.
3d Stage. *a* (Gray hepatization.) or	Aspect much distressed. Face pale and livid. Great failure of vital powers. Hectic and delirium. Cough continues, and the sputa are either absent, or sometimes they remain rust-colored; at others become like prune-juice, even fetid.	Absolute dulness on percussion. Tubular breathing and bronchophony, frequently with gurgling râles where the lung is disorganized.
b (Resolution.)	Symptoms yielding about 7th day of disease. Cough less troublesome, expectoration easier. Patient evidently convalescing.	Dulness diminishing by absorption. Broncho-vesicular breathing with *crepitant redux* râles yielding to normal vesicular murmur, and percussion-note.

POST-MORTEM APPEARANCES.—*Lungs:* 1st stage. Engorged with frothy andbloody serum. Dark-red color externally, and on section. Crepitating less, and heavier, than sound lung, but still floating in water. Pulmonary tissue slightly softened.

2d. Red externally, red or mottled and granular on cut surface, and

of liver-like solidity.   Easily torn, and with fluid exuding on pressure
less abundant than in first stage, but thicker, and towards the end of
this stage becoming purulent.   Not crepitating, and sinking in water.

3d. Reddish-yellow or gray.   More soft and friable.   Purulent
fluid exudes from the cut surface; and, on pressure, the whole lung
may be reduced to a pulp-like mass.

PLEURISY.

The symptoms of pleurisy are attributed to inflammation of the
serous covering of the lungs, and is to be distinguished from passive
effusion into the pleural sac, or *hydrothorax*, which is readily recognized by the following points of difference:

PLEURISY.

Due to inflammation (active).

Has an acute beginning, accompanied by stitch in the side, cough, constitutional disturbance, pyrexia, etc.

May be traced to traumatic causes, or to exposure to wet and cold; or may complicate zymotic diseases.

One side only affected, as a rule.

Runs its course in a few days, terminating in chronic pleuritic effusion, or in absorption of fluid.

No complications generally.

HYDROTHORAX.

Due to transudation (passive).

Effusion takes place insidiously, without local or general symptoms, beyond those caused mechanically by pressure on the thoracic viscera.

Due to blood disorder, accompanying renal disease, or more rarely to obstruction to circulation by morbid growths, or valvular disease of heart.

May be bilateral.

Remains stationary for months, or may slowly increase.

Accompanied by albuminuria, heart disease, and dropsy elsewhere in the body.

PLEURISY.

	SYMPTOMS.	PHYSICAL SIGNS.
PLEURISY: 1st Stage, or Stage of Hyperæmia.	Rigors, or more frequently mere chilliness. Sharp, stabbing pain in the side, increased by deep inspiration or cough, accompanied by more or less tenderness on pressure. Breathing short and hurried. Respiration chiefly abdominal, with inability to lie on the affected side. Short, dry cough, or none at all. Pulse full and bounding. Febrile symptoms.	Diminished movement on the affected side. Friction fremitus may sometimes be felt. Percussion sound not materially altered. Vesicular murmur feeble and jerking in rhythm. To-and-fro-friction sound.
2d Stage, or Stage of Effusion.	Cough, dyspnœa, sense of weight and fullness of the affected side. Febrile symptoms less marked. Patient lies toward, not on, the affected side. Complexion inclined to be dusky.	Almost total absence of movement of the affected side, which is unduly prominent, the intercostal spaces being obliterated or even bulging. Integuments occasionally œdematous. Vocal vibrations absent. Complete dulness on percussion, most marked in the dependent portions of the chest, and sometimes altered by change of posture. Heart pushed over to sound side, and diaphragm pushed down, so that the liver and stomach descend lower into the abdomen than in health. Vesicular murmur almost, or quite, absent. Frequently bronchial breathing near the spine. Voice sounds absent or feeble, except when the layer of fluid is thin, and then there may be ægophony. No friction sound. Puerile breathing in sound lung.

PLEURISY—(*Continued*).

	SYMPTOMS.	PHYSICAL SIGNS.
3d Stage (Resolution after Effusion).	Gradual diminution of the cough, dyspnœa, and other symptoms. Returning ability of the patient to lie on the sound side. Gradual return of displaced organs to their normal position.	The movement of the chest gradually increases. Return of vocal vibration and friction fremitus. The dulness on percussion diminishes from above downward, but the resonance generally remains box-like for a considerable period. Gradual restoration of the vesicular murmur, at first weak and distant, then somewhat harsh, and subsequently of a normal character. Reappearance of the friction sound for a time. Pseudo râles occasionally to be heard. Ægophony sometimes to be heard, more often bronchophony, and ultimately normal vocal resonance.

POST-MORTEM APPEARANCES.—1st stage. Pleura opaque and drier than natural, roughened and highly vascular, and presenting a close network of blood-vessels with ecchymoses.

2d. Fluid either serous or purulent, mixed with shreds of creamy lymph, in the cavity of the pleura. Lung pushed upward and backward towards the spine, its surface coated with a layer of lymph of the same kind as that mixed with the fluid. The lung collapsed and carnified.

3d. If the effusion has been of long duration the lung remains carnified and bound down by adhesions, and the chest-wall undergoes retraction or depression, the ribs overlap, and there is more or less lateral curvature of the dorsal spine toward the diseased, and of the lumbar toward the healthy side.

The diagnosis can be made by drawing off part of the fluid from the chest by means of a hypodermic syringe. If purulent, it should be evacuated; if allowed to remain it may lead to amyloid changes in the liver and kidneys.

The effusion may become purulent at first; there are no reliable means of recognizing the exact time when this occurs, but later it assumes all the characteristics of empyema, as follows:

	SYMPTOMS.	PHYSICAL SIGNS.
Empyema.	More decided febrile disturbance of a hectic type; night-sweats. Morning remission and evening exacerbations, Face puffy and semi-transparent. Clubbing of the finger ends. If pointing inwardly, abundant purulent sputa.	The physical signs are those of the stage of effusion, except that there is entirely absent vocal fremitus and increased resistance upon percussion. The diagnosis is often to be determined only with the aid of the aspirator, or the hypodermic needle.

DIAGNOSIS BETWEEN PLEURISY WITH EFFUSION AND PNEUMONIC CONSOLIDATION.

PLEURISY.	PNEUMONIA.
1. Begins with chilliness or several slight rigors.	1. Begins with a severe and protracted rigor.
2. Sharp, catching, stitch-like pain in the side.	2. Pain does not catch the breath; is more of a dull character.
3. Cough, dry or with little mucous expectoration, very painful, and repressed by the patient.	3. Cough frequent and severe, with rusty, viscid expectoration.
4. Pyrexia is not great and the skin may be moist.	4. Great febrile disturbance, skin hot and pungent.
5. Excretion of chlorides not affected.	5. Diminution or absence of chlorides in the urine.
6. Pulse-respiration ratio not affected, except in excessive effusion.	6. Pulse-respiration ratio may fall to 2 : 1.
7. Affected side rounded; displacement of heart.	7. No alteration in shape of the chest or of the intercostal spaces; heart not displaced.

DIAGNOSIS BETWEEN PLEURISY WITH EFFUSION AND PNEUMONIC CONSOLIDATION—(*Continued*).

PLEURISY.	PNEUMONIA.
8. Feeble or absent vocal fremitus.	8. Vocal fremitus usually much intensified.
9. Absolute dulness on percussion, transcending the median line in front.	9. Less intense dulness, not transcending the median line.
10. Feeble or absent vesicular breathing; bronchial breathing at the root of the lung.	10. Marked tubular breathing, often of a metallic character.
11. Vocal resonance absent below, sometimes ægophonic above.	11. Loud bronchophony.

DIAGNOSIS BETWEEN PNEUMONIA AND PULMONARY APOPLEXY.

PULMONARY APOPLEXY.*	PNEUMONIA.
Nearly always associated with heart disease or pyæmia. May follow traumatism.	Generally an independent disease in robust individuals. Due to infection or exposure.
Onset sudden. Fever absent, except in pyæmia. Pulse irregular and intermittent. Signs of the rupture of a vessel in the lungs.	Onset with malaise and chill. Fever. Pulse rapid.
Expectoration blackish, with small dark clots. Hemorrhage may be profuse.	Expectoration rust-colored; no clots.
Dyspnœa severe at first, afterwards diminishing.	Dyspnœa gradually grows in intensity.
Dulness distinctly circumscribed; respiration bronchial, with moist râle.	Dulness larger and extending. Crepitant râle. Tubular breathing, bronchophony.
A peculiar acid and alliaceous odor to the breath "like the smell of tincture of horse-radish" (GUENEAU DE MUSSY).	Not present.
Usually fatal if hemorrhage is large.	Prognosis usually favorable under proper treatment.

* Pulmonary infarction presents symptoms like those of pulmonary apoplexy, but less marked.

THROMBOSIS OF PULMONARY ARTERY.

The symptoms of an immediately fatal attack are : Sudden extreme dyspnœa with open tubes, cough and thoracic pain, lividity or pallor, rapidly failing pulse, cold sweats, intense anxiety, and attacks of fainting or unconsciousness, with or without spasms.

In the *diagnosis*, the suddenness of the conditions being of the chief interest, all those forms of suffocation requiring time for their production may be disregarded, and there remain :—

1. Closure of the greater air-passages or of a large number of small ones, from without or from within.

2. Nervous lesions, particularly intra-cranial, affecting respiration and circulation.

3. Obstruction to the pulmonary circulation from emboli, of blood and air particularly (fat being more gradual in its effects).

Physical and rational evidence of open air passages eliminate the first series.   In intra-cranial origins of suffocation the predominant early symptoms are those of cerebral anæmia, namely, pallor, relaxed muscles, disturbed hearing and vision, contracted pupils, fainting and convulsions.   Dyspnœa may sometimes precede these symptoms, but it is not of so severe a character as in the other series.

In favor of the third is the history of an antecedent thrombus, or of a disease of the heart likely to be associated with thrombosis, or of septicæmia.   Pulmonary thrombosis may also occur after child-birth, and be mistaken for hysteria.   Asthma presents such marked clinical features, that due attention to these will prevent any difficulty in diagnosis, except where it occurs as a complication.

ASTHMA.

SYMPTOMS.

There may be premonitory symptoms, such as gradually increasing dyspnœa or the passing of a large quantity of limpid urine; but the attacks usually come on suddenly at an early hour in the morning; the patient awakes in a start, with a sensation of suffoca-

PHYSICAL SIGNS.

Chest generally distended, though there is scarcely any expansive movement.   Recession of the intercostal spaces, supra-sternal and supra-clavicular fossæ and epigastrium during inspiration, which is short and jerky, while expiration is prolonged and wheezing.   Vocal

ASTHMA—(*Continued*).

SYMPTOMS.	PHYSICAL SIGNS.
tion and oppressiveness of the chest; he either sits upright in bed, or sometimes stands holding on to a piece of furniture, so as to bring into play the accessory muscles of respiration. Countenance pale and anxious; in bad cases cyanotic. Skin covered with sweat; extremities cold. Pulse frequently feeble. The attacks generally terminate with the expulsion of tough, ashy gray pellets of mucus.	vibration not markedly affected. Rhonchal fremitus may be felt. Resonance on percussion increased all over the chest. Almost complete absence of vesicular murmur. Every variety and kind of sibilus and rhonchus, whistling, squeaking, cooing, snoring sounds, and occasionally mucous râles towards the termination.

POST-MORTEM APPEARANCES.—The appearances found after death are principally the result of chronic bronchitis and emphysema, with dilatation of the right side of the huart.

PNEUMOTHORAX.

This condition is commonly found associated with a serous effusion —pneumo-hydrothorax; but occasionally presents itself as an independent affection. The characteristics of the two forms are as follows :

	PNEUMOTHORAX.	PNEUMO-HYDROTHORAX.
Symptoms.	Generally sharp, stabbing, pain, with the sensation of something having given way. Urgent dyspnœa and evidences of shock. More or less cyanosis. Posture assumed by patient varies. Pulse frequent, weak and small. Respiration may be 40 to 60 in the minute. Troublesome cough without expectoration. In some cases of phthisis, or where there are extensive pleural adhe-	Symptoms the same, except that the cough is usually attended by fetid, or muco-purulent expectoration: The patient lies on or toward the affected side.

PNEUMOTHORAX—(*Continued.*)

	PNEUMOTHORAX.	PNEUMO-HYDROTHORAX.
Symptoms. *Continued.*	sions, pneumothorax may come on quite imperceptibly.	
Physical Signs.	Dilatation of the affected side, with obliteration or bulging of the intercostal spaces. Movements of respiration diminished or absent. Increased elasticity of the walls of the chest. Feeble or absent vocal fremitus. Clear tympanitic resonance on percussion. If the amount of air is extreme there may be high-pitched dullness. No true vesicular murmur; bronchial breathing may be heard along the spine. Amphoric sounds, with inspiration, voice, and cough, also a metallic echo; the bell sound may be elicited. The neighboring viscera are displaced to a variable degree.	Same as in simple pneumothorax, except that percussion is dull in the lower part of the chest, and tympanitic above the level of the fluid. Metallic tinkling and splashing sounds on succussion are also frequently heard.
Post-mortem Appearance.	Lung collapsed, lying near vertebral column, unless bound down by old adhesions to some other part of the chest wall. The gas is composed chiefly of carbonic acid and nitrogen, containing but little oxygen, and occasionally some sulphureted hydrogen.	Lung collapsed. Air, mixed with fluid, in pleural cavity. Mostly arises as a termination to phthisis, a superficial cavity becoming ruptured. May occur in pneumonia, emphysema, or gangrene of the lung, and more rarely in other diseases.

EMPHYSEMA.

This affection presents itself in two forms, the vesicular and the interlobular, which are distinguished as follows:

	VESICULAR EMPHYSEMA.	INTERLOBULAR EMPHYSEMA.
Symptoms.	Habitual shortness of breath, with occasional paroxysms of urgent dyspnœa, most frequently supervening on catarrh. Cough, with or without expectoration of thin, transparent, frothy mucus. In the last stage of the disease there are symptoms due to interference with the circulation, as palpitation, cyanosis, general dropsy, and congestion of the abdominal viscera. The disorder is essentially chronic in its course, and may progress so slowly as not to materially shorten life. It generally occurs in persons who are otherwise vigorous, and is hence supposed to grant immunity from consumption.	Urgent dyspnœa and oppression, generally occurring suddenly after some violent effort of coughing, the subcutaneous areolar tissue frequently becoming œdematous.
Physical Signs.	Chest "barrel-shaped" and almost circular. Sternum projecting forward. Scapulæ and clavicles raised and ill-defined. Ribs more horizontal, and intercostal spaces widened. Respiration abdominal.	No marked change of contour of the chest. Percussion tympanitic over the affected part.

EMPHYSEMA—(*Continued*).

	VESICULAR EMPHYSEMA.	INTERLOBULAR EMPHYSEMA.
Physical Signs. *Continued.*	Movement of chest much diminished. Heart beating in the epigastric region. Resonance on percussion greatly increased or tympanitic. Feeble inspiration, prolonged expiration, the former wheezing, the latter generally with rhonchus or sibilus. Vocal fremitus and resonance usually deficient.	
Post-mortem Appearance.	Lung does not collapse as usual when the chest is opened, but, on the contrary, may rise up and bulge out of its cavity. It is pale and anæmic, and does not crepitate when pressed, but feels soft and downy, and is drier than ordinary. The air cells are dilated, or several have become one cavity from the rupture of the septa between them. Cells vary from the size of a millet-seed to that of a swan shot, or larger.	Bead-like bubbles of air seen through the pleura, or partitions between the lobules much widened. Sometimes air is found beneath the areolar tissue of the neck.

CANCER OF THE LUNG.

The principal obstacle in recognizing this disease is the liability to confound it when primary and unilateral (as it usually is when primary) with phthisis. Similar cough, emaciation, hæmoptysis, night sweats, etc., occur in both. The points of difference are :—

PULMONARY CANCER.	PHTHISIS.
Sides of chest more markedly asymmetrical; the tumor may bulge through the intercostal spaces.	One side may be sunken; never bulging.
Percussion dulness very great; may extend beyond median line.	Percussion dulness moderate; never extends beyond median line.
Frequent changes in the signs of auscultation, râles, bruits, etc.	Changes much more gradual.
Hæmoptoic sputa, " resembling currant jelly."	Sputa never present this appearance.
Pain constant, severe, lancinating.	Pain variable, intermittent.
Cancerous cachexia, tinge of skin, etc.	Absent.
Temperature may be subnormal.	Temperature usually elevated. Hectic.

Pulmonary cancer is sometimes so masked that its diagnosis requires the closest attention. It may be present without the characteristic sputa, without cachexia, and even without pain at cancerous spot.* Such instances are, of course, very rare.

It is liable to be mistaken for *chronic pleurisy* or *vice versa*. The distinguishing features are, that in cancer there is an absence of the complete consolidation of chronic pleurisy; the consolidation of the latter is at the lower portion of the lung; the expectoration of cancer is quite different from that of pleurisy and bronchitis; and the previous history, both of the individual and his family, in cancer, points to this disease, while chronic pleurisy has as an antecedent an acute attack.

The deposits of gummatous nodules in the lungs consequent on *secondary syphilis*, together with the cachexia attendant on that disease, may simulate a cancerous deposit. The history of the case, the presence of syphilitic signs in other organs and tissues, and the fact that cancers tend to spread and infiltrate the surrounding tissue, while the syphilitic nodule remains isolated and circumscribed, are the distinctive points. (For characters of syphilis of the lung see page 176.)

* See case recorded in the *Boston Medical and Surgical Journal*, January, 1876.

CHAPTER III.
DISEASES OF THE CIRCULATORY APPARATUS.

The anatomical positions of the several parts of the heart are as follows :—

RELATIONS OF THE HEART TO THE PRÆCORDIAL REGION.

REGION.	SITUATION.
APEX OF HEART . .	Ordinarily found in health between fifth and sixth ribs on left side, about two inches below the nipple and one inch on its sternal side.
BASE OF HEART. . .	On a level with the third costal cartilages.
TRICUSPID ORIFICE .	Extends from the junction of the fourth left costal cartilage with the sternum, behind that bone to the articulation of it with the sixth right cartilage.
MITRAL ORIFICE . .	To the left of the tricuspid valves, immediately behind the fourth costal cartilage; but less superficially placed than the tricuspid.

(192)

THE PRÆCORDIAL REGION—(*Continued*).

REGION.	SITUATION.
PULMONARY ORIFICE.	Immediately behind the left border of the sternum at the junction of the third costal cartilage with that bone.
AORTIC ORIFICE . .	About half an inch lower than and to the right of the pulmonary orifice, behind the sternum, on a level with the third interspace.

Let it be remembered that the tricuspid orifice is the most superficial, then the pulmonary, next the aortic, and deepest of all is the mitral orifice. Ranged from above downward, the pulmonary orifice comes first, then the aortic, then the mitral, and lastly the tricuspid.

PHYSICAL EXAMINATION OF PRÆCORDIAL REGION.

EXAMINATION BY	SHOWS.
INSPECTION	Form of chest. Point at which the apex of the heart strikes the wall of the chest. Regularity of impulse, and extent over which it is perceptible.
PALPATION	Force, extent of diffusion, and regularity of impulse. Presence or absence of purring tremor or of friction fremitus.
PERCUSSION	Extent and intensity of præcordial dulness.
AUSCULTATION . . .	Character of rhythm. Character of sounds, normal or abnormal.

THE AREA OF SUPERFICIAL CARDIAC DULNESS

Is roughly triangular in shape, the right side of the triangle being the mid-sternal line from the level of the fourth chondro-sternal articulation downward; the hypothenuse being a line drawn from the same articulation to a point immediately above the apex-beat; the base being a line drawn from immediately below the apex-beat to the point of meeting between the upper limit of liver dullness and the mid-sternal line (Dr. GEE).

13

NORMAL SOUNDS AND IMPULSE OF HEART.

SOUND.	CHARACTER.	POINT OF GREATER INTENSITY.	CAUSE.	TIME.	CONDITION OF CIRCULATION.
FIRST SOUND (Systolic).	Dull and prolonged.	Fourth and fifth intercostal spaces just within left nipple line.	Closure of auriculo-ventricular valves, and, perhaps, muscular contraction of the ventricles themselves; also impact of apex against the chest-wall, and vibration of papillary muscles and chordæ tendinæ.	$\frac{4}{10}$	Contraction of ventricles, following that of auricles. Closure of auriculo-ventricular valves, open aortic and pulmonary valves; propulsion of blood into the arteries. Impulse of the heart immediately followed by pulse at the wrist.
FIRST PAUSE.	ℓ	. . . . . .	. . . . . .	$\frac{1}{10}$	Auricles dilating.
SECOND SOUND (Diastolic).	Short and clear.	Base of heart, opposite the third right costal cartilage.	Sudden closure of the aortic and pulmonary valves.	$\frac{2}{10}$	Filling of both auricles and and ventricles. Closure of arterial valves, opening of auriculo-ventricular valves.
SECOND PAUSE.	. . . . . .	. . . . . .	. . . . . .	$\frac{3}{10}$	Complete distention of auricles, followed by their contraction, and distention of ventricles. Auriculo-ventricular valves open, arterial valves closed.
IMPULSE.	. . . . . .	Between fifth and sixth ribs on left side, about one and a half or two inches below the nipple, and one inch to its inner side.	In part due to the tilting upward of the apex, but chiefly to the recoil of heart and change in shape, for during the systole it becomes harder and more globular.		

ENDOCARDIAL MURMURS.

TIME.	SITUATION.	ORIFICE.	NATURE.
SYSTOLIC.	Basic (right).	Aortic.	Obstructive (stenosis).
	" (left).	Pulmonary.	" "
	Apical.	Mitral.	Regurgitant (insufficiency).
		Tricuspid.	" "
DIASTOLIC.	Basic.	Aortic.	" "
PRESYSTOLIC.	Apical.	Mitral.	Obstructive (stenosis).

Pulmonary regurgitant murmur (diastolic) and tricuspid obstructive murmur (presystolic) are very rarely met with clinically, and for all practical purposes they may be disregarded.

The most frequent combinations of these murmurs are :—

1. Combined aortic obstruction with regurgitation (systolic and diastolic murmur at right base).

2. Mitral obstruction and regurgitation (presystolic and systolic apical murmur).

3. Various combinations of the two preceding forms, the aortic and mitral valves being both diseased.

4. Mitral obstruction with dilated right ventricle, and consequently tricuspid regurgitation (Dr. AITKEN). (Systolic murmur heard best at lower part of sternum).

Order of frequency of endocardial murmurs, commencing with the most common :—

1. Mitral regurgitant.
2. Aortic constrictive.
3. Aortic regurgitant.
4. Mitral constrictive.
5. Tricuspid regurgitant.
6. Pulmonary constrictive.
7. Pulmonary regurgitant.
8. Tricuspid constrictive.

Order of relative gravity as "estimated not only by their ultimate lethal tendency, but by the amount of complicated miseries they inflict."—Dr. WALSHE :

1. Tricuspid regurgitation.
2. Mitral constriction and regurgitation.
3. Aortic regurgitation.
4. Pulmonary constriction.
5. Aortic constriction.

GENERAL RULES FOR THE DIAGNOSIS OF HEART DISEASE.

Dr. John Hughes Bennett * gives the following rules:

1. In health the cardiac dullness, on percussion, measures, immediately below the nipple, two inches across, and the extent of dullness beyond this measurement commonly indicates either the increased size of the organ or undue distention of the pericardium.

2. In health the apex of the heart may be felt and seen to strike the chest between the fifth and sixth ribs, a little below and a little to the inside of the left nipple. Any variations that may exist in the position of the apex are indications of disease, either of the heart itself or of the parts around it.

3. A friction murmur synchronous with the heart's movements indicates pericardial or ex-pericardial exudation.

4. A bellows murmur with the first sound, heard loudest over the apex, indicates mitral insufficiency.

5. A bellows murmur with the second sound heard loudest at the base indicates aortic insufficiency.

6. A bellows murmur with the second sound heard at the apex is rare. It indicates—1st, aortic disease, the murmur being propagated downward to the apex; or 2d, roughened auricular surface of the mitral valves; or 3d, mitral obstruction (pre-systolic murmur at apex).

7. A murmur with the first sound loudest at the base, and propagated in the direction of the large arteries, is more common. It indicates—1st, an altered condition of the blood, as in anæmia; or 2d, dilatation or disease of the aorta itself; or 3d, stricture of the aortic orifice, or disease of the aortic valve.

8. Hypertrophy of the heart may exist independent of any valvular lesion, but this is rare.

9. The pulse, as a general rule, is soft and irregular in mitral disease, but hard, jerking, or regular in aortic disease.

10. Cerebral symptoms are more marked in aortic disease; pulmonary symptoms in mitral disease.

Various constitutional symptoms should, in default of other obvious

* " Lectures on the Principles and Practice of Medicine."

causation, lead to the suspicion of disease of the heart. These are mainly:

1. *Symptoms referred to the circulation.* Violent, continued *pulsation* may arise from cardiac hypertrophy, and especially aortic regurgitation. *Cyanosis*, blueness of the lips, coldness of the finger tips, etc., are common in many cardiac cases. *Dropsy*, commencing in lower extremities, is a late and dangerous symptom.

2. *Symptoms referred to the lungs.* These are frequent cardiac complications, especially dyspnœa, orthopnœa, and cough.

3. *Symptoms referred to the brain.* Vertigo, languor, chorea, epilepsy, apoplexy, and paralysis may all be brought about by heart disease. In sudden cerebral attacks in patients suffering with valvular disease, *embolism* is often at work.

4. *Stomach symptoms.* Dyspepsia and hemorrhoids may find their origin in cardiac lesions.

5. *Throat symptoms.* Pain in the throat is complained of in angina; hoarseness and aphonia sometimes attend pericarditis.

6. *Renal symptoms* may follow heart disease. In all cases of cardiac disease the urine should be tested for albumen, as this condition may excite cardiac symptoms.*

CLUBBING OF THE FINGER ENDS IN CHRONIC HEART DISEASE AND PHTHISIS.

The following aphorisms on this point are laid down by Dr. HORACE DOBELL:†

Aphorism I. Clubbing of the finger ends on one or both sides of the body, with or without incurvations of the nails, may occur whenever the return of blood by one or both subclavian veins is seriously obstructed for a considerable length of time.

II. Symmetrical clubbing of the finger ends of both hands without incurvation of the sides and tips of the nails, is presumptive evidence of the existence of heart disease.

III. Clubbing of the finger ends without incurvature of the sides

---

* See also paper by Prof. DaCosta and Dr. Longstreth in *Am. Journal for Medical Sciences*, for July, 1880, for pathological relationship of heart disease and chronic kidney disorder.

† "Affections of the Heart." London, 1876.

and tips of the nails is presumptive evidence *against* the existence of phthisis.

IV. Symmetrical clubbing of the finger ends conjoined with incurvation of the sides and tips of the nails, is a sign that obstruction of the return blood by the subclavian veins and wasting of adipose tissue have co-existed.

DIFFERENTIAL SIGNS BETWEEN ANÆMIC AND ORGANIC CARDIAC SOUNDS.

ANÆMIC SOUNDS.	ORGANIC SOUNDS.
First sound heard over the right ventricle is distinct, second ringing; a *soft* systolic murmur is heard at left border of sternum.	Murmur generally *harsh* and blowing, and takes the place of one or both sounds of the heart. It may be distinctly located at apex or base.
Sounds vary in character, at different times, may disappear and reappear.	Sound the same after several examinations.
Sounds increase in intensity in following the aorta.	Sounds diminish in intensity in receding from the heart.
Pressure with the stethoscope increases or develops the sound.	Not affected by pressure.
Bruit du diable, a continuous musical hum, can be heard in the hollow above the right clavicle.	Not present (except when caused by pressure with the stethoscope).
Co-existence of pallor or anæmia; amenorrhœa; leucorrhœa; nervous exhaustion; chorea; renal disease; phthisis.	Co-existence of alteration in size of the heart; other organic signs; history of rheumatism.

PAIN AT THE HEART.

Pain is by no means a common symptom of heart disease. Not more than one in a dozen cases of chronic organic cardiac disease complain of pain at all.*   In acute cardiac affections it is more frequent.   In many cases of alleged pain at the heart, it will be found on examination to proceed from indigestion, myalgia, intercostal neuralgia, enlarged spleen, mediastinal growth, pleurisy, or pericarditis.

* SANSOM, " Diagnosis of Diseases of the Heart," p. 3.

APHORISMS OF DR. HORACE DOBELL.*

I. Pain in the region of the heart and down the left arm does not necessarily indicate heart disease.

II. The conjunction of pain in the region of the heart and pain in the left arm may be a most important symptom of heart disease, and is never to be disregarded.

III. If pain is excited by exercise taken when the stomach is not distended with food or gas, and especially if it comes on quickly and increases steadily in severity with the continuance of exercise, it is almost certain there is some serious disease of the circulatory organs.

IV. When it is found that flatulence or a full meal embarrasses the heart painfully, a careful investigation should be made into the condition both of the organ itself, and of the blood.

V. Important heart disease may exist, and yet pain at the heart and in its neigborhood be absent.

VI. The appalling import of *pain in the throat* in heart disease increases in proportion as the period of its onset deviates from the following order of severity:—

1. Pain under the left breast.
2. Pain extending from under the left breast to mid-sternum.
3. Pain extending from mid-sternum toward the left shoulder.
4. Pain extending from the left shoulder down the left arm.
5. Pain extending from mid-sternum toward the right shoulder.
6. Pain extending from the left shoulder down the right arm.
7. Pain extending up the sternum toward the region of the throat.
8. Pain in the thyroid cartilage.

When this order of advance is maintained as the exciting cause is continued, pain in the throat expresses the degree of dangerous persistence in the exciting cause of heart distress, rather than the degree of danger in the disease itself.

VII. In proportion as the *right* side of the chest and *right* arm take precedence in the order of extension of pain at the heart and its neighborhood, the probability increases that the aorta is more diseased than the heart.

* "On Affections of the Heart." London, 1876.

VIII. The volume of blood and other conditions being normal, the facility with which the pulse at the wrist is stopped by inspiration measures the *loss of heart power*.

ANGINA PECTORIS.

This disease is usually thought to be one typically connected with pain at the heart. This is by no means the case, as in many instances there is merely a sense of præcordial distress, but no actual pain (SANSOM). The diagnostic characters are :—

1. The attacks are paroxysmal, coming on at varying intervals and duration (from a minute to an hour), without assignable cause.

2. There is always a *sensation of coldness* experienced, and often a cold sweat.

3. The heart's action is not increased, and may be diminished.

4. The chest is fixed and breathing slow.

5. The pain, when present, may be of great intensity, of a cold, sickening character, directly referred to the heart, with an accompanying sense of impending dissolution.

Though essentially a neurosis, probably of the sympathetic (cardiac ganglia), angina pectoris is generally associated with some progressive degeneration of the muscular texture of the heart or coronary vessels.

DIFFERENTIAL SIGNS OF AORTIC OBSTRUCTION AND AORTIC INCOMPETENCY.

AORTIC OBSTRUCTION.		AORTIC INCOMPETENCY.
Hypertrophy of left ventricle.	**Effect on Heart.**	Hypertrophy and dilatation of left ventricle.
To left.	**Apex Displaced.**	Downward and to left.
To left greatly.	**Cardiac Dulness Increased.**	Downward and to left, more increased than in obstruction.
Forcible.	**Character of Impulse.**	More forcible than in obstruction, and over wider area.

DIFFERENTIAL SIGNS OF AORTIC OBSTRUCTION AND AORTIC INCOMPETENCY—(*Continued*).

AORTIC OBSTRUCTION.		AORTIC INCOMPETENCY.
To left of sternum.	**Impulse Felt.**	To left of sternum.
Onward, ventriculo-aortic.	**Murmur, its Direction.**	Backward; aortic-ventricular.
Systolic; loudest at beginning of systole.	**Time of Murmur.**	Diastolic; post-systolic; loudest at beginning of diastole.
Right border of sternum, in second intercostal space.	**Point of Greatest Intensity,**	Right border of sternum, opposite third intercostal space.
Upward to right sterno-clavicular articulation.	**Direction in which Propagated.**	Downward along sternum and toward apex.
Loud, harsh, or blowing.	**Character of Sound** (very uncertain and of little value for diagnosis).	Of higher pitch than in obstruction, and loudness decreases rapidly from commencement.
Murmur replaces first sound at base.	**Relation to Normal Heart Sounds.**	Replaces second at base, and occupies more or less of the pause.
Depends on condition of valves, but aortic second sound generally feeble.	**Effect on Second Sound.**	Apparent intensification of pulmonary second.
Systolic; in second right intercostal space.	**Thrill.**	Down sternum; diastolic.
Characteristic tracing with sphygmograph.	**Effects on Pulse.**	Visible pulsation in arteries (locomotive pulse).
Normal, or perhaps decreased.	**Frequency.**	Normal, or perhaps decreased.
Diminished.	**Volume.**	Increased.

DIFFERENTIAL SIGNS OF AORTIC OBSTRUCTION AND AORTIC INCOMPETENCY—(*Continued*).

AORTIC OBSTRUCTION.		AORTIC INCOMPETENCY.
Diminished.	**Power.**	Increased.
Regular.	**Rhythm.**	Regular.
Slow.	**Duration.**	Quick.
Arterial anæmia ; angina pectoris often present.	**General Tendency.**	As in obstruction, but sudden death more common than in any other form of valvular disease.

DIFFERENTIAL SIGNS BETWEEN MITRAL OBSTRUCTION AND MITRAL INCOMPETENCY.

MITRAL OBSTRUCTION.		MITRAL INCOMPETENCY.
Hypertrophy and dilatation of left auricle, and right chambers.	**Effect on Heart.**	Hypertrophy and dilatation of all four chambers.
To left and slightly downward.	**Apex Displaced.**	To left and downward.
To right of sternum, also to left at base, greatly.	**Cardiac Dulness Increased.**	To right of sternum, and also to left and downward.
Feeble, undulating, and diffused.	**Character of Impulse.**	Even more deficient in force.
To right of sternum and in epigastrium.	**Impulse, where found ?**	Generally increased all over cardiac region.
Onward; auriculo-ventricular.	**Murmur, its Direction.**	Backward; ventriculo-auricular.
Diastolic, presystolic, loudest at termination of diastole.	**Murmur, Time.**	Systolic, loudest at beginning of systole.
A little within and upward from apex beat.	**Point of Greatest Intensity.**	A little outward and upward from apex beat.

DIFFERENTIAL SIGNS BETWEEN MITRAL OBSTRUCTION AND MITRAL INCOMPETENCY—(*Continued*).

MITRAL OBSTRUCTION.		MITRAL INCOMPETENCY.
Upward and inward toward right base.	Direction in which Propagated.	Upward toward left base, and backward into axilla, and behind.
Generally rough and harsh.	Character of Sound (very uncertain and of little value for diagnosis).	Blowing, bellows murmur.
Precedes the first at apex, which is often very loud (presystolic.)	Relation to Normal Heart Sounds.	Replaces first at apex (systolic.)
Intensification of pulmonary second.	Effect on Second Sound.	Intensification of pulmonary second.
Presystolic; upward and inward from apex.	Thrill.	At apex and toward axilla.
	Effect on Pulse.	
Increased.	Frequency.	Increased.
Diminished.	Volume.	Somewhat diminished.
Diminished greatly.	Power.	Diminished a little.
Very irregular.	Rhythm.	Somewhat irregular.
Quick.	Duration,	Nearly normal.
Pulmonary and venous congestion and slow death by asphyxia.	General Tendency to	As in obstruction, but there is more tendency to dropsy. Death by asthenia.

DIFFERENTIAL SIGNS BETWEEN PULMONARY OB–STRUCTION AND TRICUSPID REGURGITATION.

PULMONARY OBSTRUCTION.		TRICUSPID REGURGITATION.
Systolic, onward, ventriculo-pulmonary.	Murmur. ·	Systolic, backward, ventriculo-auricular.
Left border of sternum, in second interspace.	Point of Greatest Intensity.	Base of ensiform cartilage.

DIFFERENTIAL SIGNS BETWEEN PULMONARY OB-
STRUCTION AND TRICUSPID REGURGITATION—
(*Continued*).

PULMONARY OBSTRUCTION.		TRICUSPID REGURGI-TATION.
Generally anæmia. Sometimes pressure of solidified lung (phthisical or pneumonic) upon the artery. Rarely organic, and then usually congenital.	Cause.	Generally secondary to disease of the lung or of left side of the heart.
Frequently *bruit de diable* in the jugular veins.	Associated Signs.	Systolic pulsation of the distended jugular veins.

Endocardial murmurs can be distinguished from pericardial by attention to the following physical signs :—

PERICARDITIS.

STAGE.	SYMPTOMS.	PHYSICAL SIGNS.
1st Stage. (Inflammation without effu-sion.)	If occurring during the course of acute rheumatism the disease may come on insidiously. Pain and tenderness in the cardiac region. Palpitation. Increased frequency of the pulse. Shortness of breath. Anxiety. Pyrexia.	Greater extent of visible impulse than natural, and on palpation the impulse is found to be more forcible, but unequal. Friction fremitus rare. Area of dulness not altered. Single or double friction sound, often preceded by a cantering action of the heart.* Heart sounds may be unchanged or even louder than in health, or they may be masked by the friction sounds.

* Cantering action of the heart, beside being met with in commencing pericarditis, is also caused by reduplication of the first or second sound of the heart against the thoracic wall at the moment of diastole, generally due to pericardial adhesions.

PERICARDITIS—(*Continued*).

STAGE.	SYMPTOMS.	PHYSICAL SIGNS.
2d Stage. (With effusion.)	Less pain. Pulse small, frequent, and sometimes irregular. Dyspnœa and often orthopnœa. Irritable cough. Loss of voice. Dysphagia. Fullness of veins in the neck. Duskiness of complexion. Great anxiety. Sleeplessness. Delirium.	Bulging of the præcordial region. Impulse displaced upward and outward; undulatory. On palpation, feeble and sometimes not perceptible; irregular. Area of cardiac dulness increased, first noticed at the base of the heart, and afterward extending to left of apex beat, increased by the recumbent posture. Heart sounds feeble, distant, and muffled at apex, louder and more superficial at base. Friction may or may not be heard.
3d Stage. (Resolution.)	A gradual subsidence of the symptoms of the second stage.	Diminution of the dulness from above and laterally. Heart sounds become clearer. Friction sounds may be heard with increased intensity.

POST-MORTEM APPEARANCES.—1st. Pericardium is dry, inflamed and has lost its polish. Exudation of lymph on both surfaces, but more on the visceral. The membrane may have a shaggy appearance.

2d. Fluid in variable quantity in the sac of the pericardium. Usually sero-fibrinous, containing flocculi of lymph. It may be purulent or bloody.

3d. Organized lymph on the pericardium, with or without adhesions between the two surfaces, adherent or united by mesh-like adhesions.

The Pain of Pericarditis.—Rheumatic pericarditis is more or less painful; but secondary pericarditis developing in the acute stage of infectious or the chronic period of cachectic diseases, *is invariably painless.*

Peripheric pain nearly equal on both sides of the chest; or remaining localized at the præcordial region, at the epigastrium, or at the left side

of the xyphoid cartilage, does not increase the danger of the pericarditis. But if *central*, giving rise to disturbance of circulation and respiration, and simulating that of angina pectoris, it means acute inflammation of the cardiac nerves, and marks an exceptionally bad case of
pericarditis. (Dr. WERTHEIMER, "Thèse de Paris," 1876; Dobell's
Reports.)

DIAGNOSIS BETWEEN ACUTE ENDOCARDIAL AND EXO-CARDIAL (PERICARDIAL) SOUNDS.

The sounds respectively perceptible in *endocarditis* and *pericarditis*
and allied disorders, may be discriminated by the following table :—

ENDOCARDIAL.	EXOCARDIAL.
1. A blowing sound, soft and bellows-like: not affected by pressure.	1. A creaking, rubbing, rough, to-and-fro sound, intensified by pressure of the stethoscope and by the patient bending forward.
2. A thrill may be felt on palpation.	2. On palpation, friction fremitus may be felt.
3. The sound appears distant.	3. The sound appears near.
4. May exist only with the systole or the diastole.	4. Exists with diastole as well as systole.
5. Accompanies the heart sounds.	5. Does not correspond with the rhythm of the heart.
6. Heard along the course of the great vessels, or conducted round to the back.	6. Confined to the region of the heart and limited to site of production.
7. Persistent character.	7. Rapid and frequent change in character; here to-day and gone to-morrow.
8. Area of cardiac dulness not altered.	8. Increased area of dulness, if fluid be also present.

DIFFERENTIAL SIGNS OF CARDIAC DILATATION AND PERICARDITIS WITH EFFUSION.

CARDIAC DILATATION.	PERICARDITIS WITH EFFUSION.
Dulness increased in the horizontal axis, of a square outline.	Præcordial dulness extends upward, and is of a rounded pyramidal outline, with apex above.

DIFFERENTIAL SIGNS OF CARDIAC DILATATION AND PERICARDITIS WITH EFFUSION—(*Continued*).

CARDIAC DILATATION.	PERICARDITIS WITH EFFUSION.
Heart sounds feeble but clear.	Heart sounds feeble, and distant sounding.
Transition from dulness to lung resonance more gradual.	Transition from dulness to lung resonance abrupt.
No friction sound.	Occasionally friction sound.
Limits of dulness persistent.	Limits of dulness often vary from day to day or week to week.
Apex beat felt at lower limits of cardiac dulness; impulse diffused.	Apex beat some distance above lower limit of cardiac dulness. (SANSOM.)

There is no doubt but that the general rules laid down for detecting pericardial effusion have been too vague. Dr. T. M. ROTCH, of Boston, re-examined the subject, and succeeded in fixing a more perfect diagnostic sign than any hitherto mentioned. He shows that *an area of flatness at from two to three centimetres from the right edge of the sternum in the fifth intercostal space* is almost absolutely sufficient to mark the presence of an effusion, and differentiate it from enlarged heart.*

DIFFERENTIAL SIGNS OF HYPERTROPHY AND DILATATION.

	SIMPLE HYPERTROPHY.	HYPERTROPHY WITH DILATATION.	SIMPLE DILATATION.
Palpation.	Cardiac area extended. Impulse strong, lifting, or forcing.	Extent of visible impulse greatly increased. Action regular, strong.	Extent of impulse greatly increased; but feeble, without lifting or forcing character.
Percussion.	Dulness increased laterally and downward.	Dulness lateral and downward.	Dulness increased in the horizontal axis of the heart.

* "Medical Communications of the Massachusetts Medical Society." 1878.

DIFFERENTIAL SIGNS OF HYPERTROPHY AND DILATATION—(*Continued*).

	SIMPLE HYPERTROPHY.	HYPERTROPHY WITH DILATATION.	SIMPLE DILATATION.
Auscultation.	First sound dull, prolonged, intensified; second sound intensified. No respiratory murmur over præcordium.	Both sounds prolonged.	Both sounds short, abrupt, and feeble. Feeble respiratory murmur.
Pulse.	Strong, full, incompressible.	Less strong, variable.	Weak, compressible, irregular.
General symptoms.	Fullness in the head, epigastric weight, short breath, rarely debility; Bright's disease.		Dyspnœa, cough, palpitation, portal congestion, debility, ascites.

FATTY DEGENERATION OF THE HEART.

This condition of the heart is frequently associated with dilatation. Generally the area of præcordial dullness is normal or slightly increased; the impulse weak; the apex beat indistinct; the action irregular; the first sound short and feeble; the second prolonged and intensified; pulse is irregular.

These physical signs obviously offer very little ground for a diagnosis. Of rational signs the following have been mentioned:

1. Attacks of faintness attended with sensations of great coldness, recurring without obvious cause. (DA COSTA.)

2. *Arcus senilis.* For this to be significant of cardiac degeneration, the ring must be ill-defined, rather yellowish than white, and the rest of the cornea be slightly cloudy or opaque, not clear and translucent, a tinge of jaundice being present. When this is the case, " the chances of cardiac degeneration are formidable." (SANSOM.)

3. Paroxysms of severe pain across the upper part of the sternum, and in the region of the heart.

4. Stomach derangements, accompanied sometimes by constipation, but more generally by diarrhœa and frequent vomiting.   This Dr. L. H. J. HAYNE thinks "almost pathognomonic of this disease." (*Lancet*, January, 1875.)

5. The "Cheyne-Stokes" respiration of ascending and descending rhythm is present in about one-third of the cases, and is probably dependent on atheroma of the aorta (HAYDEN).   It also occurs in disease of the medulla oblongata.   This symptom was first described in a case by Dr. CHEYNE, in 1818, as follows:

"For several days his breathing was irregular; it would entirely cease for a quarter of a minute; then it would become perceptible, though very slow; then, by degrees, it became heaving and quick; and then it would gradually cease again.   This revolution in the state of breathing occupied about a minute, during which there were about thirty acts of respiration."   In this case fatty disease of the heart was very marked, while the valves were healthy, and the aorta was "studded with steatomatous and earthy concretions."

No general attention, however, was directed to the peculiarity and striking character of this symptom, until, in 1846, Dr. STOKES urged its significance as a sign of fatty degeneration of the heart, believing that its presence was pathognomonic of this affection, and that it always berokened a fatal and not far distant termination.   That it did not necessarily depend on fatty degeneration of the heart itself, was soon shown by Dr. SEATON REID, who described a case in which the muscular structure was healthy, while the mitral and aortic valves were both incompetent, the left ventricle was hypertrophied, and the aorta dilated and atheromatous.   It remains an important and significant, if not a pathognomonic sign.

Dr. HAYDEN is of opinion that the absence of the impulse, or its extremely feeble character; the brief duration of the first sound, whether marked or sharp, in primary cases, and its almost complete or absolute extinction in those preceded by hypertrophy; the restriction of the sounds within a very limited area; and the occasional irregularity of the heart's action, will suffice, in the majority of cases, to establish the

14

diagnosis of fatty heart from the physical signs alone. He adds that the incipiency of primary fatty degeneration may be suspected, if the pulse, previously regular, becomes weak and irregular; if the surface be pale, the patient subject to dizziness or syncope, and the cardiac impulse feeble; although the sounds of the heart may not appreciably differ from their normal character.

A *slow pulse* sometimes is associated with fatty heart; but it also occurs in yellow fever, dengue, jaundice, and as a result of disorders of the vagus nerve, and follows diphtheria; or comes after an attack of malarial fever; also as a result of the administration of certain drugs, such as digitalis or aconite. In all cases it is necessary to exclude a slow pulse which is natural and peculiar to the patient. Irregular or slow pulse due to adherent pericardium may be distingushed by the history and physical signs.

CHAPTER IV.
DISEASES OF THE DIGESTIVE SYSTEM.

THE STOMACH AND BOWELS.—*Principal Symptoms—The Tongue—The Appetite—Acidity* (1) *from Fermentation*, (2) *from Hypersecretion—Pain—Flatulence—Vertigo*, (1) *Stomachal*, (2) *Cerebral—Vomiting*, (1) *Stomachal*,(2) *Cerebral—Comparison of Atonic Dyspepsia, Chronic Gastritis, Gastric Ulcer and Gastric Cancer—Indigestion or Dyspepsia—Abdominal Phthisis—Obstruction of the Bowels, Enteritis and Colitis.*

THE LIVER.—*Method of Examination—Significance of Pain in the Liver—Significance of Jaundice—Jaundice with Obstruction—Jaundice without Obstruction—Diseases Characterized by Enlargement with Smooth Surface; Enlargement with Uneven Surface; with Diminution of the Organ—Hepatic Abscess.*

INTERNAL PARASITES. — *Tape-worm — Hydatids — Round Worms — Thread Worms—Trichinosis.*

The principal symptoms to which the attention is directed in the diagnosis of diseases of the digestive organs are those connected with the tongue, the appetite, pyrosis, vomiting, flatulence, vertigo and pain.

THE TONGUE.

Late writers have shown considerable skepticism on the accuracy of the appearance of the tongue as indicative of the condition of the lining membrane of the stomach. It is true that a white and furred or a red and cracked tongue is occasionally seen in healthy subjects; but the standard of comparison should not be an ideally clean tongue, but the condition of the organ in the patient under inspection when in health. Local causes, such as carious teeth and irritating agents (tobacco, tea, mercury, etc.), must be allowed for in the examination. When these and similar considerations are weighed together with the repeated instances of simultaneous affections of the stomach and tongue

(211)

revealed by *post-mortems*, no question remains that the appearance and state of the latter organ often is of high diagnostic worth.

Dr. ROBERT FARQUHARSON states, in a lecture on the diagnosis of dyspepsia,* that in his experience the class of tongue which coincides most commonly with digestive disturbance is that in which the tongue seems to be covered with a thin, white fur, which on minute inspection is seen to be composed of a series of minute raised dots, and this usually coincides with pain immediately following meals.

If the tongue is raw and nearly stripped of epithelium, with enlarged and prominent papillæ, as we often see in phthisis, pain immediately after food and vomiting are usual symptoms, or large, red papillæ may stand in bold relief through a pale coating, or the tongue may be simply large and pale and flabby, as though too big for the mouth.

Dr. WILSON FOX specifies the following conditions of the tongue as valuable aids to diagnosis in this class of diseases:—

Dyspepsia with distinct atony of the stomach. The tongue broad, pale, and flabby, the papillæ generally enlarged, more especially on the tip and edges.

Dyspepsia from irritative causes. The tongue is redder than usual, often of a bright florid color, or even raw looking. It is often pointed at the tip, which, together with the sides, presents an extreme degree of injection, the papillæ standing out as vivid red points. This form is often associated with aphthæ, and is most common in scrofulous children and phthisical adults.

Dyspepsia from excessive or hurried eating is apt to present a tongue uniformly covered throughout the greater part of its surface with a thick fur, whitish or brownish, with some degree of enlargement and redness of the papillæ at the tip and edges.

Neuroses of the stomach display a tongue which, as a rule, is clean, though often pale, broad and flabby.

Superficial ulceration, or milk patches on the tongue occur in secondary syphilis. Psoriasis and cancer of the tongue may be distinguished by the appearance of the organ, and the presence of cachexia in the latter disease.

* *Medical Press and Circular*, July, 1877.

THE APPETITE.

Anorexia, or loss of appetite, is observed in cancer, in most inflammatory states of the stomach, in obstinate constipation, as well as in the
pyrexial state.

Boulimia, or excessive appetite, is found associated with enlargement
of the stomach, induration of its coats, also in diabetes and various
forms of mental alienation.

Capricious or depraved appetite is met with in sufferers from intestinal worms, in some cases of chronic inflammation of the stomach, as
well as in chlorosis, pregnancy and hysteria.

ACIDITY OF THE STOMACH, (1) FROM FERMENTATION (2) FROM HYPER-SECRETION.

Acidity of the stomach, pyrosis, heartburn, and water-brash, are disturbances of the digestion frequently included in one category. In all,
an excessive amount of acid is formed in the stomach; but in some
cases the origin of the acid is to be sought in fermentative action, and
in others in hyper-secretion from the coats of the stomach, thus calling
for different lines of treatment.

The following differential table, based on one given by Dr. WILSON
Fox, exhibits in a concise form the distinction between the two forms
of acidity:—

ACIDITY FROM FERMEN-TATION.	ACIDITY FROM HYPER-SECRETION.
Occurs in connection with causes which impede digestion.	Is most common as a reflex symptom, or in connection with other nervous disturbance, or with ulcer and cancer of the stomach.
Usually attains its height some hours after food, and is more marked in proportion to the size of the meal, and inversely to the digestive powers.	Occurs in the empty stomach, or rapidly after food, and is often of great intensity after a small meal.
Flatulence is common.	Flatulence is rare.
Pain not severe, and but slightly or not at all relieved by eating.	Pain more severe, most felt when the stomach is empty, and is relieved by food.

ACIDITY FROM FERMEN- TATION.	ACIDITY FROM HYPER- SECRETION.
Vomiting is rare.	Vomiting is common.
Vomited matters may contain organic acids, bacteria, torulæ and sarcinæ.	Vomited matters contain hydro-chloric acid in excess.
Urine frequently shows an alkaline reaction.	Urine rarely alkaline.

In both forms the process of digestion is impaired, but to a more marked degree in the fermentative variety, in which also, as a natural consequence, the impairment of nutrition of the patient is more obvious. As the fermentative action interferes with the functions of the liver, the stools are apt to be pale, and the patient suffer with constipation. The frequency with which attacks of gout and rheumatism are preceded by this form of acidity points to a diathetic process involving the general constitution.

PAIN.

Pain in the stomach is indicative of one of the following conditions:

1. The presence of irritating foreign bodies, as mechanical substances, corrosive poisons, blood or bile in large quantities, inflation from air or gases, etc.

2. Organic diseases altering the anatomical structure of the coats, especially gastritis, chronic ulcer, cancer, and thickening of the pylorus.

3. Perverted secretions, as in acidity.

4. Perverted innervation, which may be a local visceral neurosis,* as in forms of dyspepsia where pain is the prominent symptom, or as in cramp of the muscles of the stomach; or it may be from general disorders, as in patients of a rheumatic or gouty diathesis; or it may be referable to the general nervous system, as in pure neuralgia of the stomach and hysteria.

Pain in the stomach must be distinguished from rheumatic and other pains in the abdominal muscles immediately over the stomach. In

* Clifford Albutt's lectures on the Visceral Neuroses (London, 1884), very pointedly calls attention to a class of disorders that are often mistaken for cancer, but, unlike this, are rarely fatal, and quite amenable to treatment..

the latter the superficial tenderness is much greater; it is usually more marked in the left recti and obliqui abdominis muscles, and especially near their attachment to the ribs, where moderate pressure cannot affect the stomach, and by its independence of the digestive acts (BRIC-QUET).

Pain in the stomach is also liable to be simulated by pain in the course of the transverse colon, especially when the colon is distended with gas. The diagnosis may usually be made by gentle percussion, the note arising from tapping a distended colon being less prolonged and of a higher pitch than that elicited from the stomach. The pain from the colon is also less felt at the ensiform cartilage than in the hypochondriac regions, and often extends toward the sigmoid flexure, and is associated with other signs of intestinal flatulence.

Pain in the stomach depending on diseases of the spinal cord, is distinguished by its superficial tenderness, by the presence of other painful points in the affected nerves, and by the co-existence of other nervous, and the absence of digestive, symptoms.

Pain in the stomach may be simply neuralgic, or dependent upon gastric catarrh, or it may be due to organic lesions. The distinctive features between gastralgia and gastric ulcer are as follows : *—

NERVOUS GASTRALGIA.	ULCER OF THE STOMACH.
Pain is often independent of the ingestion of food, and may even be relieved by taking food.	Pain mostly dependent upon taking food, and its intensity varies with the quantity and the quality of the food.
Pain is often relieved by firm pressure.	Pain is increased by pressure.
Pain is rarely relieved by vomiting.	Pain after a meal usually relieved by vomiting.
Fixed points of tenderness and of subjective pain not generally present.	These are often present.
Relief is usually complete between the paroxysms.	Some pain often continues between the paroxysms.
Nutrition frequently well preserved.	Nutrition usually affected.

* W. H. Welch, Pepper's System of Medicine, vol. ii, page 516.

NERVOUS GASTRALGIA.	ULCER OF THE STOMACH.
Usually associated with other nervous affections, such as hysteria, neuralgia in other places, ovarian tenderness, etc.	Neuropathic states less constantly present.
Benefited less by regulation of diet than by electricity and tonic treatment.	Benefited not by electricity, but by regulation of diet.
Not followed by dilatation of the stomach.	Dilatation of the stomach may supervene.
No local alteration of temperature (PETER).	Surface temperature of epigastrium elevated (PETER).

FLATULENCE AND ERUCTATION.

Dyspeptics generally suffer with gases in the stomach, producing eructations. These gases are either generated from imperfectly digested food or are derived from the capillaries.

Eructations having the taste or odor of spoiled eggs, and occurring during the process of digestion, indicate the presence of sulphuretted hydrogen, from the decomposition of food.

When the eructations are odorless, and occur chiefly in an empty state of the stomach, they indicate the escape from the blood of carbonic acid, hydrogen or nitrogen, through the coats of the capillaries. (Oftener in hysterical subjects.)

In the former case the indications are to use anti-ferments; while in the latter relief is often attained by simply regulating the hours of meals, so as to avoid long intervals between the times of taking food.

GASTRIC VERTIGO. (VERTIGO E STOMACHO LÆSO.)

Stomachal vertigo may be difficult to distinguish as such, because in all vertiginous attacks, the stomach is disturbed. In undoubted examples the vertigo always bears some distinct relation to the condition of the stomach, coming on only when that organ is full, or only when it is empty, or only after certain articles of food, as shell-fish, strawberries, coffee, fresh bread, etc. There are also generally some dyspeptic symptoms other than vertigo complained of. Some other points are mentioned in the following table :—

STOMACHAL VERTIGO.	CEREBRAL VERTIGO.
Usually appears in definite relation to taking food; either after a meal, after particular ingesta, or on an empty stomach.	Occurs without relation to the taking of food.
Generally occurs in middle life.	Occurs in advanced life.
The apparent motion is felt to be subjective, not real (GOWERS).	A sense of movement or actual turning of objects.
Special senses not involved beyond perverted vision. Consciousness never lost.	Deafness and tinnitus aurium often present. Sometimes loss of consciousness.

VOMITING, (1) FROM DISEASE OF THE STOMACH, (2) FROM DISEASES OF THE BRAIN.

Persistent vomiting is a frequent symptom of obstinate gastric disturbance; and it has also been frequently noted as a symptom associated with organic diseases of the brain and cord, not unfrequently masking them and diverting the attention of the practitioner from the real seat of lesion. Thus in suddenly induced cerebral anæmia, in the commencement of the paralysis which follows diphtheria, in tubercular meningitis, in concussion of the brain, in poisoning affecting the brain and cord, and in fact in almost any disease of the cerebral centres, but especially the meninges, it is possible that one of the earlier and prominent symptoms will be obstinate vomiting.

A comparison of the leading clinical features of these two forms shows that they may be readily distinguished.

In a general way it may be stated that vomiting arising from the stomach is attended with more or less pain, with a furred tongue, with constipation or diarrhœa, sense of weight at the epigastrium, and preceded for a considerable period by a sense of nausea.

Vomiting from cerebral causes, on the other hand, is usually characterized by an absence of these symptoms, by a clean tongue, and a history of freedom from digestive disturbance.

Dr. Romberg has given the following criteria for its discrimination when the vomiting is of cerebral origin:

1. The influence of the position of the head; the vomiting is arrested in the horizontal, and recurs and is frequently repeated in the erect position.

2. The prevailing absence of premonitory nausea.

3. The peculiar character of the act of vomiting; the contents of the stomach are ejected without fatigue or retching, as the milk is rejected by babies at the breast.

4. The complication with other phenomena, the more frequent of which are pains in the head, and irregularity of the cardiac and radial pulse, increased during and subsequent to the act of vomiting.

The following differential table further exhibits the points of contrast (from Dr. W. Fox):

GASTRIC VOMITING.	CEREBRAL VOMITING.
Epigastric pain and tenderness are common, and in some cases very marked.	Epigastric tenderness and pain are rare.
Nausea is constant.	Nausea is frequently absent.
Oppression and weight at the epigastrium are constant.	These are rare.
Bowels are variable.	Bowels are constipated.
The tongue is loaded, except in certain cases of cancer or ulcer.	The tongue is usually clean.
Headache is absent, or not intense, chiefly frontal, of gradual invasion, and relieved by vomiting.	Headache often violent, the invasion sudden, and not relieved by vomiting.
Vertigo is rare and relieved by the vomiting.	Vertigo is very frequent, and not relieved by the vomiting.
Other nervous phenomena are rarely present, and then only in slighter forms, and relieved by vomiting.	Indistinctness of vision and diplopia. Confusion of ideas. Loss of memory. Not relieved by vomiting. Anæsthesia or paræsthesia, paralysis or cramp, convulsion or coma, are common or soon supervene.

The indications derived from the nature of the matters thrown up in vomiting are as follows :

Ingesta. The food is returned unaltered, or but slightly changed, in nervous vomiting; in a half digested state and strongly acid in chronic inflammation and cancer of the stomach; mixed with the microscopic forms such as sarcinæ and torulæ in chronic gastritis, gastric ulcer, and cancer.

Mucus is vomited in a catarrhal or sub-inflammatory condition of the stomach.

Bile appears whenever the retching is long and violent, and does not indicate any special disease.

Pus is not formed in the stomach, and when present in the vomit indicates disease in the œsophagus or air passages.

Fæces also indicates a disease elsewhere than the stomach, usually an obstruction of the intestinal canal.

Blood is vomited in gastric cancer and ulcer, in severe gastritis, in external injuries, vicariously (of the uterus), and frequently from disease of the heart or liver, producing distention of the capillaries. The presence of blood directly proceeding from the stomach, says Dr. Fox, if accompanied by severe pain, is almost pathognomonic of either gastric ulcer or cancer.

CHRONIC GASTRITIS, GASTRIC ULCER, AND GASTRIC CANCER.

The chief points in the diagnosis of diseases of the stomach are those connected with the differentiation of simple indigestion (atony of the stomach), inflammatory dyspepsia (gastritis, gastric catarrh, catarrhal inflammation of the stomach), gastric ulcer and gastric cancer.

From this group the nervous disturbances of the stomach are broadly marked off by the superficial character of the pain in these latter, its independence of the acts of digestion and the nature of the food, the co-existence of other neuralgia, the frequent absence of emaciation and other disturbances of nutrition, and the sex and age of the patients.

In reference to the value of *percussion* in the diagnosis of gastric cancer, Professor PETER, of Paris, has directed attention to the fact that when superficial percussion is made over the epigastric region somewhat distended by gas, there is found at certain points, especially in the region of the greater curvature, a certain obscurity of the note alternating with the zones of sonority. But this sign is absolutely wanting on deep percussion such as is ordinarily employed. Prof. PETER, by this means, detected a cancer of the stomach situated at the posterior surface of the greater curvature, with some cancerous nodules probably disseminated through the epiploon below the splenic region and also in the hypogastric region. At this last point also superficial percussion gave the same results.

An early sign of gastric cancer is the presence of enlarged glands in the *skin of the navel* (MAUNDER). To ascertain the mobility and outline of the stomach, the patient may be desired to drink one or two tumblers of soda water or a seidlitz powder. This distends the stomach and makes the tumor prominent.

The following comparative table* will be found useful in distinguishing between the grave forms of stomach disorder:

GASTRIC CANCER.	GASTRIC ULCER.	CHRONIC CATARRHAL GASTRITIS.
Tumor present in three-fourths of the cases.	Tumor rare.	No tumor.
Rare under forty years of age.	May occur at any age after childhood; one-half of cases under forty years.	May occur at any age.
Average duration about one year; rarely over two years.	Duration indefinite; may be several years.	Duration indefinite.
Hemorrhage f r e q u e n t; rarely profuse; most common in cachectic stage.	Less frequent; often profuse; early.	Gastric hemorrhage rare.
Vomiting often has the peculiarities of that of dilatation of the stomach.	Rarely referable to dilatation, and then only late in the disease.	Vomiting may or may not be present.
Free hydrochloric acid usually absent from the gastric contents in cancerous dilatation of the stomach.	Free hydrochloric acid usually present in the gastric contents.	Free hydochloric acid may be present or absent.
Cancerous fragments may be found in the washings from the stomach or in the vomit (rare).	Absent.	Absent.
Secondary cancers may be recognized in the liver, the peritoneum, the lymphatic glands, and rarely in other parts of the body.	Absent.	Absent.

* W. H. Welch, Pepper's System of Medicine, vol. ii., page 570.

GASTRIC CANCER.	GASTRIC ULCER.	CHRONIC CATARRHAL GASTRITIS.
Loss of flesh and strength, and development of cachexia usually more marked and rapid than in ulcer or gastritis, and less explicable by the gastric symptoms.	Cachectic appearances usually less marked, and of later occurrence than in cancer, and more manifestly dependent upon the gastric disorders.	When uncomplicated usually no appearance of cachexia.
Epigastric pain is often more continuous; less dependent upon taking food; less relieved by vomiting and less localized than in ulcer.	Pain is often more paroxysmal, more influenced by taking food, oftener relieved by vomiting, and more sharply localized than in cancer.	The pain or distress induced by taking food is usually less severe than in cancer or in ulcer. Fixed point of tenderness usually absent.
Causation not known.	Causation not known.	Often relerable to some known cause, such as abuse of alcohol, gormandizing, and certain diseases, such as phthisis, Bright's disease, cirrhosis of the liver, etc.
No improvement, or only temporary improvement in the course of the disease.	Sometimes a history of one or more previous similar attacks. The coma may be irregular and intermittent. Usually marked improvement by regulation of diet.	May be a history of previous similar attacks. More amenable to regulation of diet than is cancer.

INDIGESTION AND DYSPEPSIA.

Although the distinction is not generally drawn in ordinary language between dyspepsia and indigestion, it should not be forgotten that they are not synonymous. Dyspepsia has reference to an altered condition of the digestive fluid, its deficiency or excess, or to an organic affection of the muscular walls of the stomach, which has for its result imperfect or difficult chymification of the food; indigestion merely expresses a disturbance of function, and refers to the result rather than the cause. In dyspepsia the peptic glands or muscular apparatus of the stomach are defective, in indigestion they may be normal, but have their functions interfered with by improper and unaccustomed articles of food, or by reflex influence from other organs.

The symptoms of INDIGESTION are tabulated by Dr. MURCHISON as follows * :—

* "Functional Derangements of the Liver." London, 1874.

1. A feeling of weight and fulness at the epigastrium and in the region of the liver.

2. Flatulent distention of the stomach and bowels.

3. Heartburn and acid eructations.

4. A feeling of oppression, and often of weariness and aching pains in the limbs, or of insurmountable sleepiness after meals.

5. A furred tongue, which is often indented at the edges, and a clammy, bitter, metallic taste in the mouth, especially in the morning.

6. Appetite often good; at other times anorexia and nausea.

7. An excessive secretion of viscid mucus in the fauces, and at the back of the nose.

8. Constipation, the motions being scybalous, sometimes too dark, at others too light, or even clay-colored. Occasionally attacks of diarrhœa, alternating with constipation, especially if the patient be intemperate in the use of alcohol.

9. In some patients attacks of palpitation of the heart, or irregularity or intermission of the pulse.

10. In many patients occasional attacks of frontal headache.

11. In many, restlessness at night and bad dreams.

12. In some, attacks of vertigo and dimness of sight, often induced by particular articles of diet.

DYSPEPSIA may be due to impaired motion as well as to deficient secretion. The following table will give the distinctive points for diagnosis:*

	1. DYSPEPSIA FROM IMPAIRED MOTION.	2. DYSPEPSIA FROM DEFECTIVE SECRETION.
Uneasiness after meals .	Constant symptom, generally soon replaced by sense of tension accompanying flatulency.	Not infrequent, but commonly soon merged into acute pain.
Flatulence	Characteristic symptom.	Comparatively infrequent; some of the worst cases, in which pain after food and other symptoms are particularly severe, are entirely free from flatus. The tendency is to lactic, butyric, and perhaps other forms of fermentation, in which gases are not evolved.

* ARTHUR LEARED. "Dyspepsia." *British Medical Journal*, May, 1879. p. 660.

	1. DYSPEPSIA FROM IMPAIRED MOTION.	2. DYSPEPSIA FROM DEFECTIVE SECRETION.
Gastric pain.	Infrequent, but occurs occasionally, as a result of flatulence, and is peculiar in kind.	Variously described as sharp, shooting, dull, or dragging; is the most characteristic symptom of defective secretion of gastric juice.
Constipation.	Almost always a m a r k e d symptom.	Not generally present, and the bowels are in many cases relaxed.
Treatment	Strychnia, carbolic acid, thymol, charcoal.	Diet, tonics, pepsin, acids, hygienic treatment.

Hyperperistalsis, dyspepsia of fluids, flaccid stomach, and other states, give rise to similar symptoms.

CLIFFORD ALBUTT criticises the term dyspepsia, and urges the substitution, in each case, of an exact term which shall correspond with the pathology of the disorder.

ABDOMINAL PHTHISIS.

Abdominal phthisis (tubercular peritonitis), in its acute forms, closely simulates typhoid fever. There are febrile symptoms attended with remissions, heat and dryness of the surface, pains in the limbs, drowsiness and disordered secretions, diarrhœa, and emaciation. It differs from typhoid in these particulars:—

L. The pain is diffused over the abdomen, not limited to the cæcal region.

2. There are no red spots (with rare exceptions). When they occur they do not come out in crops, and are more papular and less erythematous.

3. There is generally tubercular disease in other organs.

4. The temperature may be irregularly febrile, but has not the morning remissions, and does not pursue the typical cycle of that of typhoid fever.

OBSTRUCTION OF THE BOWELS.

The causes of a mechanical stoppage of the bowels are principally the following: Intussusception; impaction of fæces; strictures, often syphilitic or cancerous; twisting of the bowel (volvulus); herniæ; pressure of tumors, and foreign bodies, such as gall-stone.

The symptom first noticed is constipation with colicky pains, which do not yield to ordinary remedies; slight distention of the abdomen, and some soreness on pressure. Vomiting follows, very severe, even becoming fecal. It is liable to be confounded with peritonitis and strangulated hernia. The following rules for diagnosis have been laid down by the eminent surgeon, Mr. JONATHAN HUTCHINSON, of London:

1. When *a child* becomes suddenly the subject of symptoms of bowel obstruction, it is probably either intussusception or peritonitis.

2. When an *elderly person* is the patient, the diagnosis will generally rest between impaction of intestinal contents and malignant disease (stricture or tumor).

3. In *middle age* the causes of obstruction may be various; but intussusception and malignant disease, both of them common at the extremes, are now very unusual.

4. Intussusception cases may be known by the frequent straining, the passage of blood and mucus, the incompleteness of the constipation, and the discovery of a sausage-like tumor, either by examination *per anum* or through the abdominal walls.

5. In intussusception, the parietes usually remain lax, and, there being but little tympanites, it is almost always possible, without much difficulty, to discover the lump (or sausage-like tumor) by manipulation under ether.

6. Malignant stricture may be suspected when, in an old person, continued abdominal uneasiness and repeated attacks of temporary constipation have preceded the illness. It is to be noted also that the constipation is often not complete.

7. If a tumor be present and pressing on the bowel, it ought to be discoverable by palpation, under ether, through the abdominal walls, or by examination by the anus or vagina, great care being taken not to be misled by scybalous masses.

8. If repeated attacks of dangerous obstruction have occurred with long intervals of perfect health, it may be suspected that the patient is the subject of a congenital diverticulum, or has bands of adhesion, or that some part of the intestine is pouched and liable to twist.

9. If, in the early part of a case, the abdomen become distended and hard, it is almost certain that there is peritonitis.

10. If the intestines continue to roll about visibly, it is almost certain that there is no peritonitis. This symptom occurs chiefly in emaciated subjects, with obstruction in the colon of long duration.

11. The tendency to vomit will usually be relative with three conditions and proportionate to them. These are (1) the nearness of the impediment to the stomach, (2) the tightness of the constriction, and (3) the persistence or otherwise with which food and medicine have been given by the mouth.

12. In case of obstruction in the colon or rectum, sickness is often wholly absent.

13. Violent retching and bile vomiting are often more troublesome in cases of gall-stones or renal calculus simulating obstruction than in true conditions of the latter.

14. Fecal vomiting can occur only when the obstruction is moderately low down. If it happen early in the case, it is a most serious symptom, as implying tightness of constriction.

15. The introduction of the fingers or entire hand into the rectum, for purposes of exploration, as recommended by Prof. SIMON, of Kiel, may often furnish useful information.

INFLAMMATORY DIARRHŒA (ENTERITIS), DYSENTERY (COLITIS), AND ENTERO-COLITIS.

These diseases, both alike in being inflammations of the mucous membrane of the intestinal tract, are frequently associated. But for therapeutic as well as prognostic purposes, it is desirable to recognize the distinctions which they present in well marked types. They are:

ENTERITIS.

Seat of inflammation is in the small intestine.

Usually begins with colic, nausea and vomiting, constipation (rarely diarrhœa), chilliness soon followed by high fever, thirst, and hot skin.

Pulse at first tense and full; soon becomes small, wiry, quick.

DYSENTERY.

Seat of inflammation is in the large intestine.

Usually begins with painless, slight diarrhœa, followed by chill, slight or no fever, sense of weight near the anus. No colic.

Pulse often a little excited; or if fever is high, full and rapid.

15

ENTERITIS.	DYSENTERY.
Pain paroxysmal, local tenderness marked, greatly increased by pressure.	Pain more moderate, usually distinctly over the colon; moderate tenderness.
Stools mucous, rarely blood, very rarely pus. No scybala. No tenesmus.	Stools scanty, bloody, contain pus, scybala, little fæces. Marked tenesmus.
Aortic pulsation felt by the patient on the right of the umbilicus.	Aortic pulsation not noticed by the patient.

Entero-colitis is very common in early childhood; it is distinguished from cholèra infantum by its inflammatory character, gradual onset, and progressive character of symptoms, and amenability to treatment, especially hygienic and dietetic.

DISEASES OF THE LIVER.

Previous to an examination of the liver, the patient should have a free action of the bowels, as fæcal accumulations are a constant cause of diagnostic errors. He should lie on his back on a firm bed, with his knees drawn up and the abdominal muscles relaxed. Palpation should be upon the patient's skin directly, not on the clothing. The physician, seating himself at the patient's right side, should apply the tips of the fingers of the right hand just below the free border of the ribs, and request the patient to make full inspiration and expiration. He will thus be able to feel the upper edge and surface of the liver and ascertain the *condition of the surface*, whether smooth or nodular. By percussion, which should be made while the patient is in the same position, the *size* of the liver can be quite accurately mapped out.

These two facts are the first steps to a diagnosis; as most hepatic diseases can be assigned to one of these classes:

1. Liver enlarged, with smooth surface.
2. Liver enlarged, with nodular surface.
3. Liver atrophied.

Pain in the hepatic region should be investigated; whether dull or acute, persistent or intermittent, etc. The condition of *jaundice* is ascertained, in light cases, by examining the under surface of the tongue and the conjunctiva of the eye, which will display the icteric discoloration when the general surface does not. A still more delicate test of the presence of jaundice is derivable from examination of the urine. The following three tests are employed by Prof. HARDY, of Paris :

1. *Chloroform.* When this is poured upon normal urine it sinks, by reason of its great density, to the bottom of the test-glass, exhibiting there a crystalline transparency. If we pour it on the icteric urine, and having shaken the test-tube plugged by the thumb, leave it quiet for a moment, the chloroform deposit contrasts strongly by its dull color with the yellow of the superficial layers—the yellow color being deeper in proportion to the quantity of bile in the urine. It is an excellent test of icteric urine.

2. *Iodine.* When the tincture of iodine is poured upon the icteric urine the mixture must not be shaken. At the upper part of the tube three very distinct colors are observable—the first layer formed by the tincture is violet; below this is a kind of diaphragm of sea-green color; and the third layer, consisting of the urine, and occupying the lowest part, is yellow.

3. *Nitric Acid.* When this agent has been poured in, the mixture, after shaking, assumes a bottle-green color passing into an olive. This is an entirely special and very characteristic appearance.*

Masset recommends the following test: The urine to be examined should be acidulated with two or three drops of concentrated sulphuric acid, and a small crystal of potassium nitrate dropped into the mixture in a test tube. A bright grass-green color will be immediately produced if the quantity of biliary coloring matter is large. Normal urine will, under the same circumstances, present a light rose-color by transmitted light.

With these hepatic symptoms determined, a study of the following tables will in most instances readily supply a correct diagnosis.

THE SIGNIFICANCE OF PAIN IN THE LIVER.

Pain having its source in the liver is divided by Dr. CHARLES MURCHISON† into three varieties, each of diagnostic significance :—

Revue de Thérapeutique, August, 1878.
† " Lectures on Diseases of the Liver."

CHARACTER OF PAIN.	DISEASES FOUND IN.
I. Pain severe, paroxysmal, with distinct intermissions; little or no local tenderness; no fever; often associated with jaundice.	Obstruction of the bile duct by gall-stones, etc. (hepatic colic); hepatic neuralgia (when jaundice is absent, probably the latter).
II. Pain moderate, continuous, slightly increased by pressure, often associated with pain in the right shoulder, slight febrile symptoms and jaundice.	Congestion and commencing inflammation of the organ; catarrh and partial obstruction of the bile ducts; acute atrophy.
III. Pain severe, constant, greatly increased by pressure, motion, coughing, etc. More or less fever; perhaps jaundice.	Always indicates inflammation of the capsule (peri-hepatitis), which may supervene in various diseases (cirrhosis, hydatids, etc.).

Hepatic pain may be simulated by various other conditions. The principal ones, with their characteristic differences, are as follows:—

1. *Pleurodynia.* The pain is strictly localized to a small spot. Absence of hepatic disturbance.

2. *Intercostal Neuralgia.* Tender points along the course of the intercostal nerve. Chiefly referred to three points in the course of the nerve: (1) The vertebral groove; (2) The axillary region; (3) The termination of the nerve in front. Co-existence of neuralgia elsewhere. Absence of hepatic symptoms. Herpes zoster will occasion no difficulty if the patient be examined with his clothing removed.

3. *Pleurisy.* Presence of pyrexia and physical signs of the disease.

4. *Gastrodynia.* Comes on with relation to food (stomach always either full or empty). Pyrosis.

5. *Intestinal Colic.* Pain referred to the umbilical region. No jaundice. Blue line of lead poisoning. Errors of diet.

6. *Renal Colic.* Pain chiefly referred to one kidney, when it shoots to the testicle and down the thigh. No jaundice. Hæmaturia and renal calculus.

Little or no hepatic pain is felt in—

1. The waxy, lardaceous, or amyloid liver.

2. The fatty liver.

3. Simple hepatic hypertrophy.

4. Hydatid tumor.

THE SIGNIFICANCE OF JAUNDICE.

The common and obvious symptom of jaundice results either (1) from obstructions of the common bile duct; or (2) independently of any obstruction of the duct. The diagnosis of these two conditions may be presented as follows:—

JAUNDICE FROM OBSTRUCTION.	JAUNDICE WITHOUT OBSTRUCTION.
When persistent, speedily becomes intense.	Persists and continues slight.
The stools are clay-colored.	The stools are natural.
Tumor in the region of the gall-bladder often present.	No tumor there.
May appear suddenly in a person in good health.	Appears gradually, unless there is a history of shock.
Intermittent jaundice in advanced life signifies gall-stones.	Intermittent jaundice in youth signifies catarrh of the duodenum.
Pain usually in severe paroxysms.	Pain usually more or less constant.
Co-existence of ascites, pregnancy, pyloric cancer (obstruction from without).	Preceding severe mental emotion, hepatic congestion, pyæmia, malarial fevers, phosphorus poisoning, epidemic prevalence, acute atrophy of liver, cancer, etc.

The principal diseases which are associated with these varieties of jaundice are the following :

JAUNDICE FROM OBSTRUCTION MAY BE DUE TO

DISEASES.	DIAGNOSIS.
1. GALL STONES.	Biliary colic present. Pain acute, paroxysmal, referred to the gall-bladder, and from this around to the right scapula. Tenderness absent or slight. Irregular rigors. No fever. Severe vomiting. Jaundice appears after a day or two. Pathognomonic sign; the presence of gall-stones in fæces.

JAUNDICE FROM OBSTRUCTION MAY BE DUE TO—
(*Continued.*)

DISEASES.	DIAGNOSIS.
2. HYDATIDS.	Liver enlarged and altered in form but painless. Biliary colic with fever, quick pulse and high temperature. Pathognomonic; hydatid vesicles in the fæces. Peculiar thrill.
3. CANCER AND TUMORS.	Antecedent history of visceral cancerous disease. Pain and nausea after taking food. A hard and sensitive tumor in the epigastric or right hypochondriac region. Hemorrhage from the stomach or bowels. Pains at night.

JAUNDICE WITHOUT OBSTRUCTION MAY BE DUE TO

DISEASE.	DIAGNOSIS.
1. MALARIAL FEVERS. Yellow Fever, Pyæmia.	History of malarial or specific poisoning, or actual presence of one of the diseases named.
2. EPIDEMIC JAUNDICE.	Gastric catarrh; stools pale; epigastric soreness; nausea or vomiting; loss of appetite; often commences with a chill after exposure. Most epidemics of jaundice seem to have been due to malarious poison or vitiated atmosphere. Infantile jaundice is of the latter character.
3. NERVOUS JAUNDICE.	History of severe mental emotion, great suffering or sudden shock. Onset rapid; often cerebral symptoms.
4. JAUNDICE FROM CONGESTION.	Feeling of weight and soreness over liver. Bad breath; poor appetite; furred tongue; vertigo. Right decubitus. Urine scanty and high colored. Slight dyspnœa. Bowels sluggish.

Acute atrophy, mineral poisons (especially by phosphorus), typhoid fever and very obstinate constipation, are other occasional causes of this form of jaundice.

CLASSIFICATION OF HEPATIC DISEASES WITH RE-GARD TO THE SIZE OF THE LIVER.*

I. LIVER ENLARGED, SURFACE SMOOTH.

SIMPLE HYPERPLASIA.	Liver enlarged, smooth, painless; absence of other symptoms.
LEUKÆMIC HYPER-PLASIA.	Liver enlarged and smooth. Spleen enlarged. Pallor of the skin. Pathognomonic; presence of a marked increase of the white blood globules, (1:20 and upward.)
CONGESTION. (a) Simple.	Enlargement moderate. Tenderness; conjunctiva jaundiced; stools pale; bowels irregular; tongue coated; low spirits; headache; vertigo; noises in the ears. No jaundice or dropsy.
(b) From cardiac disease.	Liver enlarged, smooth. Slight jaundice. Some dyspnœa. Dropsical effusions. Mitral or aortic disease. Emphysema or induration of the lungs.
(c) From malaria.	Enlargement slight. Enlarged spleen. History of malarial disease. Pathognomonic; the malarial pigment in the blood.
WAXY DEGENERATION.	Enlargement considerable, uniform, of slow growth, borders sharply defined, feel firm. Pain slight. Patient emaciated and cachectic. Splenic enlargement common. Diarrhœa and dyspepsia. History of phthisis, syphilis, or protracted suppuration.
FATTY DENGENERATION.	Enlargement considerable, borders rounded, feel doughy. No tenderness nor pain. spleen small; jaundice slight or absent. Diarrhœa. A pale, smooth, greasy skin. History of intemperance, phthisis, or indolent life.
HYDATID TUMORS.	Enlargement considerable, irregular, painless; usually of the left lobe of the organ. Feel elastic or fluctuating. Jaundice rare. Increase of size slow. No constitutional symptoms.

*Partly taken from E. J. JANEWAY, "Diagnosis of Hepatic Affections," N. Y., 1877.

CLASSIFICATION OF HEPATIC DISEASES WITH REGARD TO THE SIZE OF THE LIVER.

I. LIVER ENLARGED, SURFACE SMOOTH—(*Continued*).

SIMPLE ATROPHY.	Liver small, surface even. Preceded by ascites, dyspnœa, serious disease of heart or lungs, or signs of congestion.
ACUTE YELLOW ATRO-PHY.	Rare. Jaundice always present, though rarely intense. Pain considerable. Tenderness. Generally vomiting; splenic dulness. Pulse irregular. The typhoid state common. Urine dark, acid, sp. grav. 1.012–1.024; absence of urea, uric acid and the chlorides; presence of leucine and tyrosine (pathognomonic). Intestinal hemorrhage and hæmatemesis common.

II. LIVER ENLARGED, SURFACE NODULAR OR IRREGULAR.

ABSCESS OR TROPICAL HEPATITIS.	Liver enlarged, irregular surface bulging. Dull, heavy pain. Jaundice rare. Pyrexia and chills. History of residence in a warm climate.
CANCER.	Enlargement often very great, progressive, irregular; nodular excrescences often to be felt. Feel hard and resistant. Pain lancinating and tenderness acute. No febrile symptoms. Jaundice. "The co-existence of enlarged liver with persistent jaundice ought always to raise the suspicion of cancer" (MURCHISON). Dyspepsia, nausea, vomiting, constipation, or diarrhœa, short, dry cough, ascites. Patients over 40. In suspected cancer of the liver the urine should always be examined; half a drachm of strong nitric acid should be added to half an ounce of the urine. If the fluid changes to a dark or black hue, and especially if no albumen is present, and the liver is either increased or diminished in size, the diagnosis of *melanotic cancer* is rendered very probable. (Dr. EISELT, of Prague.)
SYPHILITIC LIVER.	Liver enlarged, surface nodulated, lobes irregular, separated by deep fissures.

CLASSIFICATION OF HEPATIC DISEASES WITH REGARD TO THE SIZE OF THE LIVER—(*Continued*).

III. LIVER DIMINISHED IN SIZE.

CIRRHOSIS, OR CHRONIC ATROPHY.	Liver small, sometimes only half size, surface granular or nodulated; "hob-nail liver." Outset insidious, with signs of disordered digestion. Dull pain and slight tenderness in hepatic region. Ascites common. Spleen often enlarged. Superficial veins of the abdomen enlarged. Hemorrhoids frequent. Jaundice rare or slight. Progressive emaciation and debility. History of spirit-drinking almost invariably.

HEPATIC ABSCESS.

It has been shown * that an obscure and chronic form of hepatic abscess is a far more common disease than is generally supposed, and that it is often exceedingly difficult of diagnosis.

These abscesses may exist without any local symptoms or such general disturbance of the system as is commonly regarded as indicating their presence, and are a very common concomitant of prolonged malarial poisoning. The pathognomonic sign of their presence is the *discovery of pus on inspiration of the parenchyma of the liver*. The place of election is one of the intercostal spaces. The rational symptoms may be collated as as follows:—

1. Gastric and intestinal derangements; dyspeptic symptoms of various kinds.

2. Slight jaundice, conjunctivæ yellow ; complexion sallow.

3. Depression of spirits, hypochondria or melancholy. This is a very usual symptom, and so important that Dr. HAMMOND recommends that in all cases of hypochondria or melancholia the region of the liver should be carefully explored, and even if no fluctuation be detected, or any other sign of abscess be discovered, aspiration, with

* TAUSCKY, *Med. Record*, April 20th, 1878; HAMMOND, St. Louis *Clin. Record*, June, 1878; BYRD, N. Y. *Med. Journal*, July, 1878, etc.

proper precautions, should be performed. If pus be evacuated, the operation may be expected to be followed by a cure of the mental disorder, as well as by the preservation of the life of the patient from the probably fatal conseqaences of hepatic abscess.

4. Sense of weight or pain in the right side; more or less tenderness on pressure (all local symptoms often absent).

5. Circumscribed fluctuation over the hepatic region. This is a positive sign, but is by no means always to be discovered.

6. Cerebral symptoms, as vertigo, cephalalgia, insomnia, hysteria, and hyperæmia.

7. Slight rigors, and feverishness, simulating some of the more chronic forms of intermittent fever.

INTESTINAL WORMS.

The symptoms to which parasites in the intestinal canal and other organs give rise are numerous, but by no means specific or definite. The following tabular arrangement sets forth the more prominent:

TAPE WORM. Tænia Solium. Tænia Saginata.	Pain and discomfort in the belly; variable appetite; constipation and diarrhœa alternating; *itching at the nose or anus* without local cause; low spirits, loss of flesh, nervous seizures. Stools variable. Pathognomonic: The discovery of scolices or joints in the stools, or about the anus, or of eggs in the fæces (microscopic).
HYDATID CYSTS. Tænia Echinococci.	These occur chiefly in the lungs and liver. (See Diseases of the Liver.) They begin with a rounded, tense, smooth, elastic swelling, painless until inflammation begins, and without other symptoms than those caused by their size. They are often attended with the "hydatid thrill." This may be felt by placing the left hand flat and closely upon the tumor, then percussing sharply with the fingers of the right hand. A long sustained tremor is observed, "like that experienced on an iron railway bridge during the passage of a train."

HYDATID CYSTS (*Continued*).

Pathognomonic: Echinococci or microscopic hydatids in the contained fluid, which may safely be drawn by aspiration.

ROUND WORMS, LUMBRICI.
Ascaris Lumbricoides.

Symptoms of intestinal irritation. Capricious appetite. Pain of a gnawing or griping character. Tenderness on deep pressure over the abdomen. Tumid condition of the belly, Alternate constipation and diarrhœa. The tongue pale, flabby, indented by the teeth, and often has a peculiar shiny appearance. Pupils generally dilated. Squinting, nervous twitchings, or even convulsions. Sleep is restless, with grating of the teeth and waking with sudden starts. Fever may appear, often of a remittent type (worm fever, verminal fever).

Pathognomonic: Worms found in alvine evacuations.

THREAD WORMS.
Oxyuris Vermicularis.

Violent itching and irritation at the anus and vagina, increased at night. Tendency to strain. Itching at the nose. Leucorrhœa.

Pathognomonic: Worms found upon examining the parts, also seen in patient's bed and his underclothing.

TRICHINOSIS.
(Trichinæ in the blood and muscular system.)
Trichina Spiralis.

First Stage: Gastro-intestinal disturbances; thirst; loss of appetite; nausea; colicky pain in the abdomen; constipation or diarrhæa; coated tongue; feverishness. *Second Stage:* Swelling and stiffness of the muscles; muscular soreness; œdema of the subcutaneous tissue; copious sweating; debility and increased fever; dyspnœa; hoarseness and loss of voice; dropsy commencing in the eyelids and face, and proceeding to the extremities; difficulty of motion and respiration.

Pathognomonic: Presence of trichinæ in the fæces, or in the muscular structure.

The differential diagnosis from *rheumatism* is in the soreness being in the muscles and not the joints; from *typhoid fever* in the unusual pain and stiffness; the early swelling, dropsy, etc.

TRICHINOSIS(*Continued*). | Trichinæ do not colonize equally through-out a muscle, but in groups here and there. It is best, therefore, to dissect out a muscle lengthwise, in order to judge of their number.

The very large number of symptoms attributed to the presence of worms in the intestinal canal is the irritation they cause, implicating the general nervous system. This, occasionally, extends so far as to produce a "worm fever," which in many respects resembles a mild remittent with unusually pronounced nervous symptoms. The tongue is pale and flabby, and often has a peculiar shiny appearance (DATE). The pupils are generally dilated. Squinting sometimes occurs, and nervous twitchings of a choreic character. The fever is often high, with great heat of skin, and the cerebral manifestations being marked, may lead to the suspicion of hydrocephalus. From this it can be distinguished by the more direct remissions; by the previous history, showing the primary symptoms to be referable to derangements of the alimentary canal; by the less obstinate constipation; and by the expulsion of worms, after a dose of calomel or castor oil.

This has also been confounded with tubercular disease. Here the most important diagnostic point is the temperature. This in tubercular disease is always high; but when the irritation is from worms it is either normal or but temporarily elevated above the normal standard. In trichinosis, however, there is usually a continued fever during the period of muscular invasion.

Prof. Da Costa and others have succeeded, by extracting a piece of muscle from a patient's arm with a peculiar shaped trocar, in discovering living trichinæ in the muscular tissue, and establishing the diagnosis beyond question.

CHAPTER V.
DISEASES OF THE URINARY ORGANS.

The Early Signs of Bright's Disease—Comparative Diagnosis of the Different Forms of Bright's Disease (Acute Parenchymatous Nephritis, Chronic Tubal Nephritis, Yellow Fatty Kidney, Secondary Contraction of Kidney, Interstitial Nephritis or Renal Cirrhosis, Albuminoid or Amyloid Renal Degeneration, Parenchymatous Renal Degeneration) —Diabetcs Mellitus and Glycosuria—Diabetes Insipidus and Hydruria —Bile in the Urine—Urinary Calculi.

General methods for the examination of the urine, and the chemical reagents and manipulations required in its analysis, are to be found in so many text-books and treatises that we may omit them here, and confine ourselves to the differential symptoms of some of the most prominent and frequent renal diseases.

THE EARLY SIGNS OF BRIGHT'S DISEASE.

The early progress of Bright's disease is often remarkably insidious, and readily escapes recognition. Nor is it always to be detected by the familiar and easy plan of testing for albumen; for this substance is by no means invariably present in the urine, even in advanced and well marked cases. FOTHERGILL justly observes that the progress of interstitial nephritis is often without the albuminous secretion for long periods.

On the other hand, it has been abundantly shown that albumen is occasionally and transiently present in the urine of persons who present no traces of nephritis; who, in fact, may be in excellent health. (Albuminuria of adolescents.)

ON SOME FORMS OF ALBUMINURIA NOT DANGEROUS TO LIFE.

The gravity of albuminuria as a symptom has been differently estimated at different times, but gradually it has become, in recent years,

(237)

to be known that albumen often appears in the urine, even in considerable quantity and very persistently, in persons free from important organic malady. Indeed, it may be maintained that some patients with persistent albuminuria are yet eligible for life-insurance at little, if at all, above ordinary rates.

It is, therefore, important to know the characteristic features of these non-dangerous albuminurias and in general the pathological import of this symptom in a given case.

Dr. GRAINGER STEWART,* recognizes the following varieties: 1, paroxysmal albuminuria; 2, dietetic albuminuria; 3, albuminuria from muscular exertion; and, 4, simple persistent albuminuria; and illustrates each with reports of cases which are markedly characteristic.

The diagnostic features of paroxysmal albuminuria are the sudden appearance of more or less albumen in the urine with numerous casts, the process lasting only a short time and recurring at intervals with or without a perceptible exciting cause. The exciting cause, according to Dr. STEWART, is irritation of the kidneys from blood-changes. The treatment should be directed, on the one hand, to the avoidance or diminution of renal irritation, and, on the other, to the regulation of the hepatic function and of the chemical processes in the body. Happily, the attacks are usually of brief duration, and he has never known them prove permanently injurious.

Dietetic albuminuria is a variety which has long been more or less distinctly recognized. Some people suffer from it whenever they indulge in certain articles of diet. In some cases one kind of food, in others many, require to be proscribed: cheese, pastry, and eggs are among the more common offenders. Of this group our present knowledge does not suffice to afford a satisfactory explanation.

The cases of albuminuria following upon muscular exertion Dr. STEWART is disposed to attribute to a general change in vascular activity. The principal indications for their treatment are met by rest, judicious diet, and attention to the general health. Those remedies which act upon the muscular fibres of the vessels deserve trial.

The features of simple persistent albuminuria are the constant presence of albumen, usually in small quantity, unattended by tube-casts,

* *American Journal of the Medical Sciences*, Jan., 1887.

diminution of urea, increased muscular tension, cardiac hypertrophy, or other consequence of renal malady, persisting for a period of months or years, and little influenced by diet or exercise.

Dr. Stewart concludes his study with a consideration of the prognosis of these groups, placing special dependence upon the quantity of the urea, the presence or absence of tube-casts, and the condition of the pulse and heart.

From the foregoing the necessity of determining the character of the form of renal disease becomes manifest. The presence of *tube-casts* has been urged as pathognomonic of renal hyperæmia and inflammation.* These must be sought for with considerable care, as from the transparent character of some of them, and the fact that they do not form a sediment, they are readily overlooked. The directions given for their search are that the urine to be examined is placed in a tall, conical glass; after three to six hours it is inspected; from the visible deposits, whether floating or sedimentary, with the pipette a quantity is taken sufficient to fill a concave slide or a shallow cell.

This little pool is first searched with a four-tenths objective, and in a little time any cast or other microscopic object it contains is found. A more careful observation is made of the object thus found with the one-fifth. When the examination of deposits has been made in this way, the conical glass of urine should be set aside (a little chloral may be added, to prevent decomposition), and after twelve hours more the examination should be repeated. Of course, it will be remembered that the hyaline cast may be found when the condition of the kidney is only one of transient hyperæmia. The character of cast is of considerable importance in prognosis. Tyson † has declared that *per se* he attaches little consequence to the presence of casts unless they are fatty or contain oil drops.

We shall now proceed to classify the diagnostic points in the differentiation of the seven forms into which the varieties of Bright's disease are now divided, premising that more than one form may exist in the same patient. (Nephritis is not a synonym for Bright's disease, since it may occur in only one kidney.—Da Costa.)

* Dr. B. A. Segur, Proceedings of the Medical Society of Kings Co., 1878, p. 241.

† Prof. Jas. Tyson, Proceedings of Phila. County Medical Society, vol. iv., p. 133.

COMPARISON OF THE DIFFERENT

	ACUTE DESQUAMATIVE NEPHRITIS.	CHRONIC TUBAL NEPHRITIS.	YELLOW FATTY KIDNEY.
History.	Sudden onset after scarlet fever or exposure to wet and cold; Œdema of the face the sign first noticed; headache, feverishness, pain in the loins, gastric disturbance.	Symptoms of more than six weeks' duration. Often history of acute nephritis. Uræmic symptoms; abnormally low temperature. Serous inflammations. Cardiac hypertrophy.	Often follows alcoholism.
Appearance.	Dropsical, more or less swollen about the face ; skin generally dry.	More or less œdema, and general anasarca. A pale, almost characteristic, waxy look.	Dropsy considerable and persistent ; renal cachexia often marked.
Urine	Scanty, smoke colored, dark when acid, red if alkalized. Highly albuminous. Specific gravity high, 1.025 – 1.030. Reddish brown sediment of epithelial, blood, and hyaline casts.	Generally scanty, though variable. Pale, albumen about one-fourth, specific gravity low, 1.005–1.015; white sediment of hyaline and epithelial casts. No blood casts.	Scanty, pale, low specific gravity, with abundant sediment of oily casts and cells filled with oil. Albumen abundant.
Prognosis.	Recovery frequent. May lead to chronic tubal nephritis.	Recovery not likely.	Almost certainly fatal.
Pathology.	Kidneys enlarged, congested, vascular ; cortical substance increased. Tubules dark and dense.	Kidney enlarged, cortical substance increased, capsules easily separated. "Large white kidney."	Kidneys enlarged, fatty, mottled, the tubes full of fat and oil cells.

FORMS OF BRIGHT'S DISEASE.

SECONDARY CONTRACTION OF KIDNEY.	INTERSTITIAL NEPHRITIS. RENAL CIRRHOSIS.	ALBUMINOID OR AMYLOID RENAL DEGENERATION.	PARENCHYMATOUS RENAL DEGENERATION.
Symptoms of more than a year's duration. Headache. Coma or convulsions. Cardiac hypertrophy. Epistaxis.	Symptoms few and faint. Often the arthritic diathesis. Exposure to cold · and fatigue. Sense of weariness. Frequent headache. Amaurosis. Cardiac hypertrophy.	Antecedent syphilis, phthisis, or osseous disease and chronic suppuration. Enlarged liver or spleen. Chronic diarrhœa.	Pregnancy diphtheria, or acute fever.
Generally some dropsy, but not very extensive. Face sallow.	Little or no dropsy. Nerve implications, as paralysis, loss of sight or hearing, etc.	Dropsy generally amenable to treatment. Emaciation. Face sallow or pallid. Dyspnœa.	Generally no dropsy.
Scanty, pale, specific gravity about 1.015. Albumen moderate. Sediment of pale casts, dark granular, fatty cells, and waxy products.	Largely increased, pale; albumen trifling; sediment little, of finely granular casts, or minute oil drops. Specific gravity low.	Largely increased. (50–60 oz.) pale or golden; albumen considerable, perhaps one-half. Specific gravity 1.007–1.015; little or no sediment; casts hyaline and waxy.	Normal in amount. Albumen $\frac{1}{16}$ to $\frac{1}{4}$ bulk.
Generally fatal, but of slow progress.	With care, not immediately dangerous, but predisposes to uræmic attacks from exposure.	Incurable, though the patient may live for years.	Recovery frequent.
Kidneys contracted, dense, capsule adherent; atrophy of the tubules.	Kidneys at first enlarged, later contracted; connective tissue increased; capsule adherent, diminished, and corrugated. " Chronically contracted " kidney.	Kidney enlarged, smooth, waxy looking.	Kidney enlarged, the parenchyma more or less hypertrophied.

16

The effort has also been made to call in the aid of the ophthalmoscope. The presence of minute white exudations in the retina, principally around the maculæ luteæ, are believed to point to the presence
of Bright's disease, and to be found in its early stages (retinitis albu-
minurica). The appearance of the retinâ in these cases is characteristic. It consists in the grouping of small white spots, the outline of
each being clearly defined ; they are invariably circular,.of extremely
small dimensions, and present the appearances of a pearl of an in-
tensely bright color, and stand out from the retina in a marked manner. The grouping of the spots is symmetrical in each eye, and is
generally in the form of a crescent. Often the urine will only yield
signs of the minutest quantities of albumen—sometimes none at all;
but hyaline casts and these white spots may be detected by the processes here described.

In the form of *amyloid degeneration* the difficulties of diagnosis are
considerable, as not only has it been generally recognized that albumen may be absent for considerable periods while the disease is
steadily advancing, but it has been abundantly shown that it may never
appear at all in fatal cases.*

It seems, therefore, certain that we possess at present no sure diagnostic sign of amyloid degeneration of the renal vessels; that on the
one hand, it is likely to be confounded with, or mistaken for, chronic
parenchymatous nephritis asising under etiological conditions ; on the
other, it runs a great risk of being altogether overlooked. But both
of these evils may be avoided with a little care. BARTELS points out
that the differential diagnosis between amyloid disease and chronic
parenchymatous nephritis depends upon the distinguishing characters
of the urine, which, in the former, is clear, with little sediment and few
casts, mostly hyaline, and scarcely ever blood-corpuscles ; in the latter
it is always more or less turbid, with considerable sediment, is dirty
colored, contains many casts of every variety, and not uncommonly
blood-corpuscles. In those cases in which no albumen was present,
there have been signs of amyloid disease in other organs; and, in order to escape error, it will be well enough to know that the absence of

* LECORCHÉ, " Maladies des Reins," Paris, 1875; LITTEN, *Berliner Klin. Wochenschrift.*

albumen from the urine does not exclude a slight degree of amyloid disease of the kidneys.

Cystic kidney is not considered worthy of special remark, since ordinary cysts are not to be recognized with any certainty during life, nor can they always be distinguished from the chronic varieties of Bright's Disease, in which they frequently are developed (DA COSTA).

DIABETES MELLITUS AND GLYCOSURIA.

The presence of *sugar in the urine* is characteristic of both these conditions. The most convenient simple test is caustic potash (MOORE's test), either in solution or small fragments. Heated with urine containing sugar, this substance immediately produces a more or less yellow or brown color, the intensity of which is in proportion to the quantity of sugar present, and a peculiar sweet smell (melassic acid). Picric acid solution with potassa and diabetic urine strikes a crimson color when heat is applied (ROBERTS).

The test usually preferred is TROMMER's or FEHLING's, which depends upon the reduction of a salt of copper by the sugar. [The Fehling's test may be obtained in a solid form from Wyeth, of Philadelphia, for the extemporaneous preparation of the solution.] Certain medicinal substances reduce copper from the test solution—as turpentine, chloroform, benzoic acid, salicylic acid, camphor, copaiba, cubebs —and if the patient is taking these, resort should be had to the fermentation test with compressed yeast or the polariscope.

Apart from this test, the presence of sugar in the urine is revealed by many indications. We may often recognize it by grayish patches on clothing or linen, which are reduced to powder when scratched with the nail. Another circumstance indicating the sugary savor of the urine, especially in the country, is the great number of flies or ants that will be attracted around the vessel containing it. Pruritus vulvæ is often caused by diabetic irritation.

The presence of sugar once determined, it remains to decide whether it arises from simple glycosuria, which is a comparatively common and not dangerous condition, or from confirmed diabetes, which is more rare and a very perilous affection. This distinction has been insisted upon by M. GÉRIN ROZES, and most clinical teachers. The contrasting features of the two disorders may be presented as follows:—

·DIABETES MELLITUS.	SIMPLE GLYCOSURIA.
Onset gradual; occurs at all ages, and without reference to known predisposing causes.	Onset sudden; more common in the aged; in persons consuming saccharine food; in the insane; in those taking chloral; in the paroxysms of ague; after sudden excitement; blows on the head; cerebral affections.*
The amount of sugar varies very little. Specific gravity of the urine high (1025–1030).	The amount of sugar varies greatly from day to day (pathognomonic, ROZES). Specific gravity not far from normal.
The absence of saccharine food makes little or no change in the urine.	The withdrawal of saccharine food diminishes the sugar.
Volumetric analysis by Fehling's method is easy.	Such analysis is obscure, owing to the quantity of creatinine and similar substances present.
Polyuria, polyphagia, polydypsia, and impotence common and well marked.	All these may be, and generally are, absent, or slightly marked.
Nervous complications frequent.	Rare.
Treatment of little avail; result usually fatal.	Treatment efficient; result usually favorable.

With the knowledge of the very fatal character of diabetes mellitus, a recognition of is earliest symptoms becomes of immense importance for treatment. Its invasion is seldom sudden, and at the very outset may be curable, which it rarely or ever is when once developed.

Various nervous symptoms are among the earliest noted, and it is a wise rule in all nervous disorders of a doubtful character to examine the urine for sugar. Changes in the character of an individual, an abnormal irritability of temper, insomnia, and extreme feeling of fatigue, *disorders of vision*, itching of the skin, pruritus of the genital organs, especially the vulva, and more or less protracted headache, are often premonitory symptoms. Intense and obstinate neuralgic pains, with-

* In a clinical lecture by DA COSTA, a case of cerebral syphilis is reported in which glycosuria occurred.—*Philadelphia Medical Times*, Jan. 22, 1887.

out obvious cause, especially in the feet and leg, should lead to the suspicion of diabetes. Recurrent boils and carbuncles are well known to accompany the diabetic condition.

Genital impotence is one of the first signs of approaching diabetes; and whenever individuals are met with who, previously virile, become weak and impotent without coinciding disease, especially of the spinal marrow, diabetes will usually be found to be the cause. Valuable information is derivable from the mouth; for besides the insatiable thirst and dry mouth, some patients complain of a disagreeable taste, which is sometimes acrid, and at others faint, or bitter, or sugary; and it is this perverted taste which contributes to maintain the thirst.

The mouth frequently exhibits an aphthous condition, while the edges and tip, and even the whole surface of the tongue, may present a red aspect, as if the aphthæ had been removed. The gums, also, are often softened, fungous or bleeding; while in some the teeth become loose, or fall out without being decayed, and in others become carious. The breath is frequently of a bad, acid smell, and the saliva, in its reaction, is acid instead of neutral. Another fact which has sometimes led to the diagnosis is the existence of intertrigo at the commissure of the lips. This intertrigo labialis is not exclusively connected with diabetes, but when met with should always lead to an examination of the urine.

With regard to the digestive organs, boulimia on the one hand, and a complete repugnance for food on the other, with dyspepsia, should lead us to suspect diabetes. The unusual thirst of diabetics prompts them to drink large quantities of water at night, and such a habit should suggest strict inquiry for other symptoms. As a general rule it may be said that whenever there is muscular debility, emaciation and anæmia, without discoverable local cause, the urine should be examined, and will almost always be found to contain either sugar or albumen.

The prognosis in a case of Diabetes Mellitus improves with the age of the patient; occurring in elderly persons, with ordinary care, it does not appear to shorten life (DA COSTA).

DIABETES INSIPIDUS (POLYURIA) AND HYDRURIA.

The habitual discharge of an excessive amount of urine of low specific gravity, and containing neither albumen nor sugar, if accompanied with progressive emaciation, excessive thirst, and loss of vital power, constitutes *diabetes insipidus;* but under various conditions excessive diuresis may be temporarily present, as in hysteria and other cerebro-spinal and nervous affections, without serious general symptoms, and constitute the condition of *hydruria.* The distinction between the two can be made by noting the coincident disease in the latter form, the slight direct impairment of the general health, the varying amount of urine voided, and by the fact that the quantity, although large, never attains those extraordinary measures—thirty to fifty pints daily—which marked cases of diabetes insipidus present. A large amount of urine is discharged by patients with amyloid degeneration of the kidney.

BILE IN URINE.

The significance of bile in urine is the same as that of jaundice, as it indicates the presence of bile in the blood (see page 229). The tests are those for the bile pigment and those for the biliary salts. The color test usually employed is that of Gmelin; a few drops of urine are placed upon a white plate and nitric acid dropped at its side; i bile pigment be present a play of colors, from grass-green to red, is produced. The same may be obtained by adding sulphuric acid to urine in a test-tube, and dropping in a crystal of potassium nitrate. The tests for the biliary salts are so complicated that they are entirely unreliable, as generally applied. For cautions and directions for their use the reader is referred to NEUBAUER and VOGEL's "Chemistry of the Urine."

REMARKS ON URINE TESTING.

Physical Characters.—In the first place, in order to have any scientific value, the examination must be made of a specimen of the whole urine, or complete amount discharged in the twenty-four hours. The bladder should be emptied at a certain hour, and all the urine (including that passed at the same hour the following day) collected and measured. In summer time the receiver should be kept in a cool place during this time, in order to prevent bacterial development and decomposition. In cases requiring extraordinary care, the urine obtained at each act of micturition may be kept by itself. This not only will give information as to the frequency of the act and the working capacity of the bladder, but will also enable a separate analysis to be made of the *urina sanguinis*, *potûs*, and *cibi*, the importance of which is insisted upon by GOLDING BIRD and others, and which we cannot dwell upon. The odor, color, consistency, and chemical composition of the urine are affected by the diet. It is well known that albumen and sugar in some cases are only present after meals, and that the prognosis of such a case is better than when either continues during fasting. The reaction and specific gravity are also different in the urine passed before from that passed after eating.

With regard to the determination of the specific gravity, the customary method yields information of no real value. In the first place, the urinometers used in hospitals and by physicians are often unreliable; and in the second place, the object of the examination is not attainable by the usual procedure. The specific gravity is desired only in order to determine the amount of solids present in solution in the urine, from which to discover whether the excretion by the kidneys of nitrogeneous waste is sufficient, in excess, or diminished. This can only be determined by an examination of a specimen of the mixed urine of the entire twenty-four hours. If carefully observed with a correct urinometer (such as Squibb's), or, more carefully still, with the picnometer and balance, the approximate amount can be readily determined. (By multiplying the last two figures by HAESER's co-efficient 2.33, which gives the proportion of urea in 1,000 parts of urine, from which the entire daily quantity discharged may be estimated.) Before taking the specific gravity the urine should be boiled and filtered while hot; if there is a deposit of urates on cooling, an equal volume of distilled water may be added to the urine, and the proper correction made. The proportion of urea may be determined more accurately by the hypobromite process, with Greene's apparatus, or its recent modification by Marshall.*

* *Zeitschrift fur Physiologische Chemie*, Vol. xi., p. 179.

The mere presence of albumen in a specimen of urine is of small importance when compared with a diminution of urea. Some years ago Dr. Wm. L. Richardson pointed out (in a communication to the American Gynæcological Society) that in pregnancy the existence of albuminuria may be practically ignored as long as the kidneys discharge the normal quantity of urea; when they fail, however, and the quantity is far below the normal, the patient is in danger of uræmic convulsions.

The detection of albumen, however, may be useful in another direction, since it furnishes valuable information indirectly by leading us to examine the urine for tube-casts. But just as albuminuria may exist without renal disease, so may tube-casts be present in the urine without indicating the presence of organic kidney affection. Neither albumen nor tube-casts, singly or in combination, will invariably determine the existence of Bright's disease.* An exception to this statement, however, is necessary, in the case of fatty and lardaceous casts.

Albuminuria may be a part of a general pyrexia (due to the effect of increased temperature upon the filtration of albumen; †) and Grainger Stewart‡ has recently published a very interesting paper on some forms of albuminuria not dangerous to life, which he divides into four classes: (1) Paroxysmal albuminuria; (2) Dietetic albuminuria; (3) Albuminuria from muscular exertion; and (4) Simple persistent albuminuria. To these, additions might yet be made without impairing the force of his statement that "some patients with persistent albuminuria are yet eligible for life insurance at little, if at all, above the ordinary rates."

On Testing for Glucose.—In conclusion, there are several fallacies in the test for for glucose which are sometimes overlooked. Patients who are taking chloral, chloroform, benzoic or salicylic acid, turpentine, copaiba, cubebs, or camphor, and other medicinal substances, will fur-

---

* Prof. James Tyson. Proceedings of the Philadelphia County Medical Society, Vol. iv., pp. 133, 134.

† A Loewy. Auf die Einfluss der Temperatur auf die Filtration von Eiweiss-lösungen durch Thierische Membranen.—*Zeitschrift für Physiologische Chemie*, Vol. ix., H. 9, 1885. Also editorial *Philadelphia Medical Times*, Vol. xvi., p. 18.

‡ *The American Journal of the Medical Sciences*, January, 1887.

nish urine producing a deposit of the oxide of copper by Fehling's test. Indeed, the Fehling's solution itself should first be tested with some dilute normal urine in order to see that it is reliable, since it readily undergoes changes that lead it to spontaneously deposit the copper. A recent case, which occurred in Baltimore, is very instructive. An applicant for life insurance was rejected by several companies on the ground that he had a high grade of diabetes mellitus. Upon applying the copper test, an abundant orange-colored precipitate was formed, beyond question. Dr. T. B. BRUNE, of Baltimore, found that the fermentation test and polariscope both failed to give evidence of the presence of sugar.* Prof. TYSON was called in consultation, and decided that the reducing substance was not glucose. The urine was subsequently tested by Prof. WORMLEY and Dr. MARSHALL at the University of Pennsylvania, who believed the substance to be an acid not recognized heretofore. It has been suggested that this new reducing substance is oxaluric acid, which had been previously found by SCHUNCK.†

URINARY CALCULI.

There are but three forms of calculi which are at all of common occurrence, and which are, therefore, likely to demand analysis. These are *uric acid and its compounds, oxalate of lime*, and the *mixed phosphates*. Calculi of *xanthine* and *cystine* are found, though very rarely.

1. *Uric acid calculi* are the most common. They are either red or some shade of red, and usually smooth, but may be tuberculated. They leave a mere trace of residue after ignition.

2. *Oxalate of lime calculi* are frequently met with. They are generally of a dark brown or dark gray color, and from their frequently tuberculated surface have been called mulberry calculi. They may, however, also be smooth. Considerable residue remains after ignition. The calculus is soluble in mineral acids without effervescence.

3. *Calculi of the mixed phosphates or fusible calculi* are composed of the phosphate of lime and of the triple phosphate of ammonia and magnesia. They form the external layer of many calculi of different

* A reducing substance in urine resembling glucose.—*Boston Medical and Surgical Journal*, Vol. cxv., p. 621.

† *Medical Register*, Vol. i., p. 10.

composition, and may form entire calculi, but very seldom form the nuclei of other calculi.   They are white, exceeeingly brittle, fuse in the blow-pipe flame, and are soluble in acids, but insoluble in alkalies.

Few calculi of large size are of the same composition throughout. Oxalate of lime is the most frequent nucleus; uric acid may also serve as a nucleus, but phosphates, as stated, almost never.   Small collections of organic matter, as blood-clots, frequently form nuclei, and may often be recognized by the odor of ammonia on ignition. It is not uncommon to find calculi made up of concentric layers of different composition.

TO DETERMINE THE COMPOSITION OF CALCULI.*

Heat a portion of the *powdered* calculus to redness upon platinum foil.   Note whether there is a residue.

A. There is a fixed residue.   To a portion of the original powder apply the murexid test.   (This is as follows: Dissolve a small portion of the powder in a drop of nitric acid on a porcelain plate, then carefully evaporate over a spirit lamp.   When dry add a drop or two of liquor ammoniæ, when, if uric acid is present, a beautiful *purple color* will appear where the ammonia spreads.)

I. A purple color results; *uric acid* is present.   Observe whether a portion of the calculus melts on being heated.

    a. It melts and communicates—

        1. A strong yellow color to the flame of a spirit lamp; *sodium urate.*

        2. A violet color to the flame; *potassium urate.*

    b. It does not melt.   Dissolve the residue after ignition in a little dilute HCl, add ammonia until alkaline, and then ammonium carbonate solution.

        1. A white precipitate falls; *calcium urate.*

        2. No precipitate.   Add some hydric sodic phosphate solution; a white crystalline precipitate falls; *magnesium urate.*

* The processes here given are taken, with slight alterations, from Thudichum's work on the *Pathology of the Urine.*

II. No purple color results. Observe whether a portion of the calculus melts on being heated strongly.

    a. It melts (fusible calculus). Treat the residue with acetic acid; it dissolves. Add to the solution ammonia in excess; a white crystalline precipitate falls; *ammonio-magnesium phosphate.* In case the melted residue is insoluble in acetic acid, treat with HCl; it dissolves. Add to the solution ammonia; a white precipitate indicates *calcium phosphate.*

    b. It does not melt. Moisten the residue with water, and test its reaction with litmus paper; it is not alkaline. Treat with HCl; it dissolves without effervescence. Add to the solution ammonia in excess; white precipitate; *calcium phosphate.* Treat the calculus with acetic acid; it does not dissolve. Treat the residue after heating with acetic acid; it dissolves with effervescence; *calcium oxalate.* Treat the original calculus with acetic acid; it dissolves with effervescence; *calcium carbonate.*

B. There is no fixed residue. Apply the murexid test (p. 250).

I. A purple color is developed.

    a. Mix a portion of the powdered calculus with a little lime and moisten with a little water; ammonia is evolved, and a red litmus paper suspended over the mass is turned blue; *ammonium urate.*

    b. No ammonia; *uric acid.*

II. No purple color.

    a. But the nitric acid solution turns yellow as it is evaporated, and leaves a residue insoluble in potassium carbonate; *xanthine.*

    b. The nitric acid solution turns dark brown, and leaves a residue soluble in ammonia; *cystine.*

INDEX.